STUDY GUIDE with SELECTED SOLUTIONS

CALCULUS
&ITS APPLICATIONS
SEVENTH EDITION

BRIEF CALCULUS
AND ITS APPLICATIONS
SEVENTH EDITION

STUDY GUIDE with SELECTED SOLUTIONS

CALCULUS
&ITS APPLICATIONS
SEVENTH EDITION

BRIEF CALCULUS
AND ITS APPLICATIONS
SEVENTH EDITION

GOLDSTEIN • LAY • SCHNEIDER

PRENTICE HALL, Upper Saddle River, NJ 07458

Production Editor: *Carole Suraci*
Production Supervisor: *Joan Eurell*
Acquisitions Editor: *George Lobell*
Supplements Editor: *Audra Walsh*
Production Coordinator: *Alan Fischer*

©1996 by Prentice-Hall, Inc.
Simon & Schuster / A Viacom Company
Upper Saddle River, New Jersey 07458

Printed in the United States of America

10 9 8 7 6 5 4 3 2

ISBN: 0-13-375205-4

Prentice-Hall International (UK) Limited, *London*
Prentice-Hall of Australia Pty. Limited, *Sydney*
Prentice-Hall Canada Inc., *Toronto*
Prentice-Hall Hispanoamericana, S.A., *Mexico*
Prentice-Hall of India Private Limited, *New Delhi*
Prentice-Hall of Japan, Inc., *Tokyo*
Simon & Schuster Asia Pte. Ltd., *Singapore*
Editora Prentice-Hall do Brasil, Ltda., *Rio de Janeiro*

CONTENTS

CONTENTS

PREFACE

Proper use of this study guide can help to improve your performance in your calculus course. There are detailed explanations here to guide you carefully and patiently through hundreds of exercises. And there are helpful hints and strategies for studying that several thousand of our students have used successfully over the past twenty years. Students who apply the techniques in this guide consistently score from five to fifteen percentage points higher on campus-wide final exams than students who have no systematic strategy for mastering the material in this course.

The study guide is organized by the sections in *Calculus and Its Applications, Seventh Edition*, by Goldstein, Lay, and Schneider. Most sections begin with some remarks about the section, its general purpose and layout, or its relation to other material. Following that, there are complete solutions to every sixth problem. Essentially all of the basic skills are covered in these problems. The solutions are written in the style of the examples in the text, but they tend to contain more warnings and side remarks than in the text. At the end of many chapters you will find a brief review to help you prepare for an exam.

Special note for readers of *Brief Calculus and Its Applications, Seventh Edition:* The material here for Chapters 0 to 5 and Chapters 7 and 8 exactly matches your text. So does the material for Sections 6.1 to 6.5. After Section 6.5, you'll find special sections marked "Brief Edition" that correspond to the rest of your Chapter 6.

The way you use this study guide should be determined by an overall strategy for success in your calculus course. We have listed below five things you can do to improve your performance and make the course an enjoyable experience. Hardly a semester passes without several of our students writing on their (anonymous) course evaluations that they never thought they would enjoy a math course, and yet they found themselves liking this course more than most of the courses in their major. If you have the proper background, there is no reason why you cannot do well in this course, and enjoy it at the same time!

1. **Work on exercises nearly every day, keeping up with the lectures in your class.** You will minimize the time you need to learn calculus if you heed this advice. Concepts in calculus are usually introduced and explained in terms of concepts you have studied earlier. When you fall behind, the class explanations will be harder to understand and you will spend more time on your own trying to catch up. In this sense, studying calculus is like learning a foreign language, where each lesson builds on the previous lessons, and if you miss one lesson you may not understand a thing that is said in class.

2. **Read the text and examples carefully before you attempt the exercises.** Then try the exercises. If you get "stuck" on a problem, look for a relevant example, but don't reread the whole example. Read just enough to get you started on the exercise. If you studied the example earlier, parts of the explanation may still be floating in your mind, and the quick glance at the example may be all you need. If the problem is complex and you get stuck again, take another peek at the example. This process of trying to recall an example that you have already studied is the key to learning how to work the problem yourself.

Many students read the examples after they have tried (unsuccessfully) to work an exercise. When a problem seems difficult, a student may look for a similar example and "copy" the example with the numbers changed. In this approach very little learning takes place. *Being able to read an explanation and understand each step is entirely different from really knowing how to work the problem by yourself.* We cannot emphasize this too strongly. You must first *attempt* a problem by yourself, with no text or study guide to help you. When you need help, accept only enough assistance to get started on the problem.

3. **Treat this study guide as a tutor.** One function of a tutor is to point out where you are making mistakes. After you work a problem whose solution is in the study guide, compare your solution to the one printed here. If your final answer is correct, have you included all the important steps? (We have included nearly all algebraic steps. You may feel comfortable with fewer steps, but be careful. Check with your instructor if you're not sure about how much detail is desired.) If your final answer is incorrect, find your first mistake and correct it. If the solution is long, then you may uncover other mistakes by reworking the part that followed your first mistake.

Another function of a tutor is to help you on a problem when you cannot proceed further. If you get stuck on a problem that has a solution in this study guide, and if the examples in the text are not sufficient, then read the first part of the solution. But only read enough to get started. *Don't sit back and watch your "tutor" work the entire problem.* You must have the practice of working it yourself. Return to the solution later for more help, if necessary.

4. Keep a list of places where errors are likely to occur. The text warns of some potential errors. This study guide will point out many more. Identify *your* weaknesses. As you work exercises, make a note of the most common mistakes you make. Do the same with any quizzes you take. Then review the list before each exam.

5. Develop a strategy for taking each test. You should be able to walk into an exam knowing what most of the questions will concern and knowing where your own strengths and weaknesses lie.

a. Read over the entire exam for a few minutes instead of beginning work immediately on the first problem. Then as you work one problem, your subconscious mind can be thinking about related problems that appear elsewhere on the exam. (This strategy works for math exams just as well as it does for essay exams, once you learn how to do it.)

b. Begin the exam by working on the problem in which you have the most confidence. Successful completion of one or two problems will help to calm the "test anxiety" (or panic!) that most people feel to some extent.

c. Move from easier to harder problems. Avoid getting bogged down in a problem where you spend a lot of time and yet have the potential to earn only a few points.

There are other suggestions we could give, but that's probably enough for now. We'll be adding to this list at strategic points in the study guide.

A Warning

Although this study guide has the potential to help you tremendously, it also has the power to undermine your chances of success. Because the guide contains complete solutions to many homework problems, you will be tempted to read the explanations here instead of working the problems yourself. While this may shorten the time you spend on homework, it will have a *disastrous* effect on your exam performance! In the long run, a proper use of the guide *will* save you time and it *will* make the time you spend more productive, but you must use the guide wisely.

CHAPTER 0

FUNCTIONS

0.1 Functions and Their Graphs

You should read Sections 0.1 and 0.2 even if you don't plan to work the exercises. The concept of a function is a fundamental idea and the notation for functions is used in nearly every section of the text.

1. The notation [-1,4] is equivalent to $-1 \leq x \leq 4$. Both -1 and 4 belong to the interval.

7. The inequalities $2 \leq x < 3$ describe the half-open interval [2,3). The right parenthesis indicates that 3 is not in the set.

13. If $f(x) = x^2 - 3x$, then $f(0) = (0)^2 - 3(0) = 0$,
$$f(5) = (5)^2 - 3(5) = 25 - 15 = 10,$$
$$f(3) = (3)^2 - 3(3) = 9 - 9 = 0,$$
$$f(-7) = (-7)^2 - 3(-7) = 49 + 21 = 70.$$

19. If $f(x) = x^2 - 2x$, then
$$f(a+1) = (a+1)^2 - 2(a+1) = (a+1)(a+1) - 2a - 2$$
$$= (a^2+2a+1) - 2a - 2 = a^2 - 1.$$
$$f(a+2) = (a+2)^2 - 2(a+2) = (a+2)(a+2) - 2a - 4$$
$$= (a^2+4a+4) - 2a - 4 = a^2 + 2a.$$

25. The domain of $g(x) = \dfrac{1}{\sqrt{3-x}}$ consists of those x for which $x < 3$, since division by zero is not permissible and since square roots of negative numbers are not defined.

31. This is not the graph of a function by the "vertical line test".

37. By referring to the graph, we find that $f(4)$ is positive since the graph of the function is above the x-axis when $x = 4$.

43. The concentration of the drug when $t = 1$ is .03 units because the graph tells us that $f(1) = .03$.

49. If the point $\left(\frac{1}{2}, \frac{2}{5}\right)$ is on the graph of the function $g(x) = \frac{3x - 1}{x^2 + 1}$, then we must have $g\left(\frac{1}{2}\right) = \frac{2}{5}$. This is the case since

$$g\left(\frac{1}{2}\right) = \frac{3\left(\frac{1}{2}\right) - 1}{\left(\frac{1}{2}\right)^2 + 1} = \frac{\frac{3}{2} - 1}{\frac{1}{4} + 1} = \frac{\frac{1}{2}}{\frac{5}{4}} = \frac{1}{2} \cdot \frac{4}{5} = \frac{2}{5}$$

so $\left(\frac{1}{2}, \frac{2}{5}\right)$ is on the graph.

55. Since $f(x) = \pi x^2$ for $x < 2$, $f(1) = \pi(1)^2 = \pi$.
Since $f(x) = 1 + x$ for $2 \leq x \leq 2.5$, $f(2) = 1 + 2 = 3$.
Since $f(x) = 4x$ for $2.5 < x$, $f(3) = 4(3) = 12$.

0.2 Some Important Functions

The most important functions here are the linear functions, the quadratic functions, and the power functions. Absolute values appear only briefly in Sections 4.5, 6.1, and then more frequently in Chapters 9 and 10.

1. The function $f(x) = 2x - 1$ is linear. Since a line is determined by any two of its points, we may choose any two points on the graph of $f(x)$ and draw the line through them. For instance, we find $f(0) = 2(0) - 1 = 0 - 1 = -1$ and $f(1) = 2(1) - 1 = 2 - 1 = 1$. So $(0,-1)$ and $(1,1)$ are two points on the graph of $f(x)$. See the sketch in the answer section of the text.

7. To find the y-intercept evaluate $f(x)$ at $x = 0$.

$$f(0) = 9(0) + 3 = 0 + 3 = 3.$$

So the y-intercept is $(0,3)$. To find the x-intercept set $f(x) = 0$ and solve for x.

$$9x + 3 = 0,$$
$$9x = -3,$$
$$x = \frac{-3}{9} = \frac{-1}{3}.$$

So the x-intercept is $\left(\frac{-1}{3}, 0\right)$.

13. **a.** We wish to write $f(x) = .2x + 50$ in the form

$f(x) = \left(\frac{K}{V}\right)x + \frac{1}{V}$, so we set

$\frac{K}{V} = .2$ and $\frac{1}{V} = 50$.

If $\frac{1}{V} = 50$, then $V = \frac{1}{50}$. Substituting $\frac{1}{50}$ for V in $\frac{K}{V} = .2$, we get

$$\frac{K}{\left(\frac{1}{50}\right)} = .2, \text{ and hence } K = .2\left(\frac{1}{50}\right) = \frac{2}{500} = \frac{1}{250}.$$

b. To find the y-intercept evaluate the function at $x = 0$.

$$y = \left(\frac{K}{V}\right)(0) + \frac{1}{V} = \frac{1}{V}.$$

The y-intercept is $\left(0, \frac{1}{V}\right)$.

To find the x-intercept set $y = 0$ and solve for x:

$$\left(\frac{K}{V}\right)x + \frac{1}{V} = 0, \qquad \left(\frac{K}{V}\right)x = -\frac{1}{V}, \qquad \text{and} \quad x = \left(\frac{V}{K}\right)\left(-\frac{1}{V}\right) = -\frac{1}{K}.$$

The x-intercept is $\left(-\frac{1}{K}, 0\right)$.

19. $f(x) = \frac{50x}{105-x}$, $0 \le x \le 100$,

$f(70) = 100$ (million dollars) from Example 5,

$f(75) = \frac{50(75)}{105-75} = \frac{50(75)}{30} = 125$ (million dollars).

Thus the added cost to remove another 5% is:

$f(75) - f(70) = 125 - 100$ (million dollars)
$= 25$ million dollars.

From Example 5 the cost of removing the final 5% of the pollutant is 525 million dollars. This is twenty-one times the cost of removing an extra 5% of pollutant after 70% has been removed.

25. If we write $y = 1 - x^2$ in the form $y = ax^2 + bx + c$, we have $y = (-1)x^2 + (0)x + 1$, so $a = -1$, $b = 0$, and $c = 1$.

31. This function is defined by three distinct linear functions. For $0 \leq x < 2$ we have $f(x) = 4 - x$. This graph is determined by two points, say at $x = 0$ and $x = 2$.

$$f(0) = 4 - 0 = 4, \quad f(2) = 4 - 2 = 2$$

So $(0,4)$ and $(2,2)$ determine this part of the graph. Draw a line segment between these points. $((0,4)$ is part of the graph but $(2,2)$ cannot yet be counted as part of the graph since $x < 2$.)

For $2 \leq x < 3$ we have $f(x) = 2x - 2$.

This graph is also determined by two points, say at $x = 2$ and $x = 3$.

$$f(2) = 2(2) - 2 = 2, \quad f(3) = 2(3) - 2 = 4.$$

So $(2,2)$ and $(3,4)$ determine this part of the graph. Draw a line segment between these points. Now $(2,2)$ is part of the graph but $(3,4)$ cannot yet be counted as part of the graph since $x < 3$.

For $x \geq 3$ we have $f(x) = x + 1$.

This graph is also determined by two points say at $x = 3$ and $x = 4$:

$$f(3) = 3 + 1 = 4, \quad f(4) = 4 + 1 = 5.$$

So $(3,4)$ and $(4,5)$ determine this part of the graph.

Draw a line segment through these points and extend it to the right as x tends to infinity.

37. $f(x) = |x|$, where

$$|x| = \begin{cases} x \text{ if } x \text{ is positive or zero,} \\ -x \text{ if } x \text{ is negative.} \end{cases}$$

Here $x = -2.5$, which is a negative number so $|x| = -(-2.5) = 2.5$. This gives $f(x) = 2.5$ when $x = -2.5$.

0.3   The Algebra of Functions

The algebraic skills in this section will be used frequently throughout the course. You should review the material here on composition of functions before you read Section 3.2.

1. $f(x) + g(x) = (x^2 + 1) + 9x = x^2 + 9x + 1$.

7. $f(x) + g(x) = \dfrac{2}{x - 3} + \dfrac{1}{x + 2}$. In order to add two fractions, the denominators must be the same. A common denom-
inator for $\dfrac{2}{x - 3}$ and $\dfrac{1}{x + 2}$ is $(x-3)(x+2)$. If we multiply
$\dfrac{2}{x - 3}$ by $\dfrac{x + 2}{x + 2}$ we obtain an equivalent expression whose
denominator is $(x-3)(x+2)$. Similarly, if we multiply
$\dfrac{1}{x + 2}$ by $\dfrac{x - 3}{x - 3}$ we obtain an equivalent expression whose
denominator is $(x-3)(x+2)$. Thus

$$f(x) + g(x) = \frac{2}{x - 3} + \frac{1}{x + 2}$$

$$= \frac{2}{x - 3} \cdot \frac{x + 2}{x + 2} + \frac{1}{x + 2} \cdot \frac{x - 3}{x - 3}$$

$$= \frac{2(x + 2)}{(x - 3)(x + 2)} + \frac{x - 3}{(x - 3)(x + 2)}$$

$$= \frac{2x + 4 + x - 3}{(x - 3)(x + 2)} = \frac{3x + 1}{(x - 3)(x + 2)} = \frac{3x + 1}{x^2 - x - 6}.$$

13. $f(x) - g(x) = \dfrac{x}{x - 2} - \dfrac{5 - x}{5 + x}$. A common denominator for
$\dfrac{x}{x - 2}$ and $\dfrac{5 - x}{5 + x}$ is $(x-2)(5+x)$. Proceeding as in exercise
7, we have

$$f(x) - g(x) = \frac{x}{x - 2} \cdot \frac{5 + x}{5 + x} - \frac{5 - x}{5 + x} \cdot \frac{x - 2}{x - 2}$$

$$= \frac{x(5 + x)}{(x - 2)(5 + x)} - \frac{(5 - x)(x - 2)}{(x - 2)(5 + x)}$$

$$= \frac{5x + x^2 - (-x^2 + 7x - 10)}{(x - 2)(5 + x)}$$

$$= \frac{2x^2 - 2x + 10}{x^2 + 3x - 10}.$$

19. $f(x+1)g(x+1) = \left[\dfrac{x + 1}{(x + 1) - 2} \right]\left[\dfrac{5 - (x + 1)}{5 + (x + 1)} \right]$

$$= \left[\frac{x + 1}{x - 1} \right]\left[\frac{4 - x}{6 + x} \right] = \frac{(x + 1)(4 - x)}{(x - 1)(6 + x)}$$

$$= \frac{-x^2 + 3x + 4}{x^2 + 5x - 6}.$$

25. To find $f(g(x))$ we substitute $g(x)$ in place of each x in

$$f(x) = x^6. \text{ Thus } f(x) = \left(g(x)\right)^6 = \left(\frac{x}{1-x}\right)^6.$$

31. If $f(x) = x$ then

$$\begin{aligned}
f(x+h) - f(x) &= (x+h)^2 - x^2 \\
&= x^2 + 2hx + h^2 - x^2 \\
&= 2hx + h^2.
\end{aligned}$$

37. $f(x) = \frac{1}{8} x$, $g(x) = 8x + 1$. Substitute $g(x)$ in place of each x in $f(x)$. Thus

$$\begin{aligned}
h(x) = f(g(x)) &= \frac{1}{8} g(x) \\
&= \frac{1}{8} [8x + 1] \\
&= x + \frac{1}{8}.
\end{aligned}$$

$g(x)$ converts British sizes to French sizes.
$f(x)$ converts French sizes to American sizes.
So $h(x) = f(g(x))$ converts British sizes to American sizes. For example, consider the British hat size of $6\frac{3}{4}$. Since $h(x) = x + \frac{1}{8}$, the American size is $h(6\frac{3}{4}) = 6\frac{3}{4} + \frac{1}{8}$ $= 6\frac{7}{8}$, which agrees with the table.

0.4 Zeros of Functions—The Quadratic Formula and Factoring

The material in this section will be used routinely through-out the text, beginning in Section 2.3. (Factoring skill is also needed for a few exercises in Section 1.4.) By Section 2.3 you must have *mastered* the ability to factor quadratic polynomials and to solve problems such as Exercises 33-38 (in Section 0.4). You should also be able to factor cubic poly-nomials where every term contains a power of x. (See (b) and (c) of Example 6.) Problems such as those in Exercises 25-32 will appear in Section 6.4 and in later sections. The quadra-tic formula should be memorized although you will not need to use it as often as factoring skills.

1. In the equation $2x^2 - 7x + 6 = 0$, we have $a = 2$, $b = -7$, and $c = 6$. Substituting these values into the quadratic formula, we find that

$$x = \frac{-(-7) \pm \sqrt{(-7)^2 - 4(2)(6)}}{2(2)} = \frac{7 \pm \sqrt{49 - 48}}{4} = \frac{7 \pm 1}{4}.$$

Thus $x = \frac{7 + 1}{4} = 2$ or $x = \frac{7 - 1}{4} = \frac{3}{2}$. So the zeros of the function $2x^2 - 7x + 6$ are 2 and $\frac{3}{2}$.

7. In the equation $5x^2 - 4x - 1 = 0$, we have $a = 5$, $b = -4$, and $c = -1$. By the quadratic formula,

$$x = \frac{-(-4) \pm \sqrt{(-4)^2 - 4(5)(-1)}}{2(5)} = \frac{4 \pm \sqrt{16 + 20}}{10}$$

$$= \frac{4 \pm \sqrt{36}}{10} = \frac{4 \pm 6}{10}.$$

So the solutions of the equation $5x^2 - 4x - 1 = 0$ are

$$x = \frac{4 + 6}{10} = \frac{10}{10} = 1, \text{ and } x = \frac{4 - 6}{10} = \frac{-2}{10} = -\frac{1}{5}.$$

13. In order to factor $x^2 + 8x + 15$ we must find values for c and d such that $cd = 15$ and $c + d = 8$. The solution is $c = 3$, $d = 5$, and

$$x^2 + 8x + 15 = (x+3)(x+5).$$

19. In order to factor $30 - 4x - 2x^2$ we first factor out the coefficient -2 in front of x^2. That is,

$$30 - 4x - 2x^2 = -2(x^2 + 2x - 15).$$

Now to factor $x^2 + 2x - 15$, we must find values for c and d such that $cd = -15$ and $c + d = 2$. The solution is $c = 5$, $d = -3$, and $x^2 + 2x - 15 = (x+5)(x-3)$. Thus

$$30 - 4x - 2x^2 = -2(x+5)(x-3).$$

25. If a point (x,y) is on both graphs, then its coordinates must satisfy both equations. That is, x and y must satisfy $y = 2x^2 - 5x - 6$ and $y = 3x + 4$. Equating the two expressions for y, we have

$$2x^2 - 5x - 6 = 3x + 4.$$

To use the quadratic formula we rewrite the equation in the form

$$2x^2 - 8x - 10 = 0.$$

Dividing both sides by 2 we obtain

$$x^2 - 4x - 5 = 0.$$

By the quadratic formula

$$x = \frac{-(-4) \pm \sqrt{(-4)^2 - 4(1)(-5)}}{2(1)} = \frac{4 \pm \sqrt{16 + 20}}{2}$$

$$= \frac{4 \pm \sqrt{36}}{2} = \frac{4 \pm 6}{2}.$$

So $x = \dfrac{4 + 6}{2} = 5$ or $x = \dfrac{4 - 6}{2} = -1$. Thus the x-coordinates of the points of intersection are 5 and -1. To find the y-coordinates, we substitute these values of x into either equation, $y = 2x^2 - 5x - 6$, or $y = 3x + 4$. Since $y = 3x + 4$ is simpler we use this equation to find that at $x = 5$, $y = 3(5) + 4 = 15 + 4 = 19$. Also, at $x = -1$, $y = 3(-1) + 4 = -3 + 4 = 1$. Thus the points of intersection are (5,19) and (-1,1).

31. As in Exercise 5, equate the two expressions for y to get

$$\tfrac{1}{2}x^3 + x^2 + 5 = 3x^2 - \tfrac{1}{2}x + 5.$$

Then rewrite this as

$$\tfrac{1}{2}x^3 - 2x^2 + \tfrac{1}{2}x = 0,$$

and factor out a common factor of x to get

$$x\left(\tfrac{1}{2}x^2 - 2x + \tfrac{1}{2}\right) = 0.$$

This tells us that one of the points of intersection has x-coordinate 0. To find the x-coordinate of the remaining points of intersection we apply the quadratic formula to $\tfrac{1}{2}x^2 - 2x + \tfrac{1}{2} = 0$:

$$x = \frac{-(-2) \pm \sqrt{(-2)^2 - 4\left(\tfrac{1}{2}\right)\left(\tfrac{1}{2}\right)}}{2\left(\tfrac{1}{2}\right)} = \frac{2 \pm \sqrt{4 - 1}}{1} = 2 \pm \sqrt{3}.$$

Substituting these values into $y = 3x^2 - \tfrac{1}{2}x + 5$, we find at $x = 0$, $y = 3(0)^2 - \tfrac{1}{2}(0) + 5 = 5$; at $x = 2 + \sqrt{3}$,

$$y = 3(2+\sqrt{3})^2 - \frac{1}{2}(2+\sqrt{3}) + 5$$

$$= 3(7+4\sqrt{3}) - \frac{1}{2}(2+\sqrt{3}) + 5$$

$$= 21 + 12\sqrt{3} - 1 - \frac{1}{2}\sqrt{3} + 5$$

$$= 25 + \frac{23}{2}\sqrt{3};$$

and at $x = 2 - \sqrt{3}$,

$$y = 3(2-\sqrt{3})^2 - \frac{1}{2}(2-\sqrt{3}) + 5$$

$$= 3(7-4\sqrt{3}) - 1 + \frac{1}{2}\sqrt{3} + 5$$

$$= 21 - 12\sqrt{3} + 4 + \frac{1}{2}\sqrt{3}$$

$$= 25 - \frac{23}{2}\sqrt{3}.$$

Thus the points of intersection are:

$$(0,5), \quad (2+\sqrt{3}, 25+\frac{23}{2}\sqrt{3}), \quad \text{and} \quad (2-\sqrt{3}, 25-\frac{23}{2}\sqrt{3}).$$

37. A rational function will be zero only if the numerator is zero. Thus we solve

$$x^2 + 14x + 49 = 0,$$
$$(x + 7)^2 = 0,$$
$$x + 7 = 0.$$

That is, $x = -7$. Since the denominator is not 0 at $x = -7$, we conclude that $x = -7$ is the solution.

0.5 Exponents and Power Functions

We have found that operations with exponents cause our students more difficulty than any other algebraic skill, so we have included lots of drill exercises. Try some of each group of problems. If you cannot work them accurately with relative ease and confidence, keep working more problems. If you need more practice, get a college algebra text or Schaum's *Outline of College Algebra*. Do this immediately, because operations with exponents are used in Section 1.3 and in most sections thereafter.

Helpful Hint: We are all so accustomed to reading formulas from left to right that it is more difficult to use a formula such as $b^r b^s = b^{r+s}$ in "reverse," that is, in the form $b^{r+s} = b^r b^s$. Yet the laws of exponents are sometimes needed in reverse from the way they are written in the text. So *memorize* the following list, as well as the list on page 48 of the text.

Laws of Exponents (continued)

1'. $b^{r+s} = b^r \cdot b^s$ 4'. $b^{rs} = (b^r)^s$

2'. $\dfrac{1}{b^r} = b^{-r}$ 5'. $a^r b^r = (ab)^r$

3'. $b^{r-s} = b^r \cdot b^{-s} = \dfrac{b^r}{b^s}$ 6'. $\dfrac{a^r}{b^r} = \left(\dfrac{a}{b}\right)^r$

A common use of Law 4' is in the form $b^{m/n} = (b^{1/n})^m$. For instance, $9^{3/2} = (9^{1/2})^3 = (3)^3 = 27$, and $27^{4/3} = (27^{1/3})^4 = (3)^4 = 81$. Most instructors will assume that you can compute the square roots of the following numbers without using a calculator:

$$4,\ 9,\ 16,\ 25,\ 36,\ 49,\ 64,\ 81,\ 100.$$

You should also know the following cube roots:

$$8^{1/3} = 2,\ 27^{1/3} = 3,\ 64^{1/3} = 4, \text{ and possibly } 125^{1/3} = 5.$$

It wouldn't hurt also to learn the following fourth roots:

$$16^{1/4} = 2,\ 81^{1/4} = 3.$$

Solutions for Section 0.5:

1. $3^3 = 3 \cdot (3 \cdot 3) = 3 \cdot 9 = 27.$

7. $-4^2 = -16.$ Note that the exponent 2 does not act on the negative sign. That is, -4^2 is not the same as $(-4)^2 = (-4) \cdot (-4) = +16.$

13. $6^{-1} = \dfrac{1}{6}$ (Law 2).

19. $(25)^{3/2} = (25^{1/2})^3$ (Law 4')

$\qquad = 5^3 = 125.$

25. $4^{-1/2} = \dfrac{1}{4^{1/2}}$ (Law 2)

$\qquad = \dfrac{1}{2}.$

31. $6^{1/3} \cdot 6^{2/3} = 6^{1/3+2/3}$ (Law 1)

$\qquad = 6^1 = 6.$

37. $\left(\dfrac{8}{27}\right)^{2/3} = \dfrac{8^{2/3}}{27^{2/3}}$ (Law 6)

$\qquad = \dfrac{(8^{1/3})^2}{(27^{1/3})^2}$ (Law 4')

$\qquad = \dfrac{2^2}{3^2} = \dfrac{4}{9}.$

43. $\dfrac{x^4 \cdot y^5}{xy^2} = \left(\dfrac{x^4}{x}\right)\left(\dfrac{y^5}{y^2}\right)$

$\qquad = (x^{4-1})(y^{5-2})$ (Law 3)

$\qquad = x^3 y^3.$

49. $(x^3 y^5)^4 = (x^3)^4 (y^5)^4$ (Law 5)

$\qquad = x^{12} y^{20}$ (Law 4).

55. $\dfrac{-x^3 y}{-xy} = \dfrac{x^3 y}{xy} = \dfrac{x^3}{x} \cdot \dfrac{y}{y}$

$\qquad = x^{3-1} y^{1-1}$ (Law 3)

$\qquad = x^2 y^0 = x^2.$

61. $\left(\dfrac{3x^2}{2y}\right)^3 = \dfrac{(3x^2)^3}{(2y)^3}$ (Law 6) $\begin{bmatrix}\text{Don't forget} \\ \text{parentheses.}\end{bmatrix}$

$\quad\quad = \dfrac{3^3(x^2)^3}{2^3 y^3}$ (Law 5) $\begin{bmatrix}\text{Careful: This step} \\ \text{is where mistakes} \\ \text{usually happen.}\end{bmatrix}$

$\quad\quad = \dfrac{27(x^2)^3}{8y^3} = \dfrac{27x^6}{8y^3}$ (Law 4).

67. $\sqrt{x}\left(\dfrac{1}{4x}\right)^{5/2} = x^{1/2}\left(\dfrac{1}{4x}\right)^{5/2}$

$\quad\quad = x^{1/2}\cdot\dfrac{1^{5/2}}{(4x)^{5/2}}$ (Law 6) $\begin{bmatrix}\text{Careful: Don't} \\ \text{forget parenthe-} \\ \text{ses around 4x.}\end{bmatrix}$

$\quad\quad = x^{1/2}\cdot\dfrac{1}{4^{5/2}x^{5/2}}$ (Law 5)

$\quad\quad = \dfrac{x^{1/2}}{(4^{1/2})^5 x^{5/2}}$ (Law 4)

$\quad\quad = \dfrac{x^{1/2}}{2^5 x^{5/2}} = \dfrac{1}{32}\cdot x^{1/2-5/2}$ (Law 3)

$\quad\quad = \dfrac{1}{32}\cdot x^{-2} = \dfrac{1}{32x^2}.$

73. $x^{-1/4} + 6x^{1/4} = x^{-1/4}(1 + 6x^{1/2}) = x^{-1/4}(1 + 6\sqrt{x}).$

79. $f(4) = (4)^{-1} = \frac{1}{4}.$

85. $A = P(1 + \frac{r}{m})^{mt}$, where $P = 100$, $r = .06$, $m = 1$, $t = 6$

$\quad = 500(1 + \frac{.06}{1})^{1\cdot6}$

$\quad = 500(1.06)^6 = \$709.26.$

91. $A = P(1 + \frac{r}{m})^{mt}$, where $P = 1500$, $r = .06$, $m = 360$, $t = 1$

$\quad = 1500(1 + \frac{.06}{360})^{360\cdot1}$

$\quad = 1500(1.0001667)^{360} = \$1592.75.$

97. [new stopping distance] $= \frac{1}{20}(2x)^2 = \frac{1}{20} \cdot 4x^2 = 4\left(\frac{1}{20}x^2\right)$

$$= 4 \cdot [\text{old stopping distance}]$$

0.6 Functions and Graphs in Applications

The crucial step in many of these problems is to express the function in one variable using the information given. Exercise 7 is a typical example. Both the height and width are expressed in terms of x. The function, here perimeter, is now in terms of x.

1. If x = width, then the height is 3(width) = 3x.

7. Perimeter = $2 \cdot$ height + $2 \cdot$ width
$$= 2(3x) + 2(x) = 6x + 2x = 8x.$$

    Area = height $\times$ width
$$= (3x)(x) = 3x^2.$$

But Area = 25 square feet, so
$$3x^2 = 25.$$

13. Volume = $\pi r^2 h$, where r is the radius of the circular ends and h is the height of the cylinder. Since the volume is supposed to be 100 cubic inches,
$$\pi r^2 h = 100.$$

Using the solution to Practice Problems 0.6,

      Area of the left end: πr^2,

      Area of the right end: πr^2,

      Area of the side (a "rolled-up rectangle"): $2\pi rh$.

Cost for the left end: $5(\pi r^2) = 5\pi r^2$,

Cost for the right end: $6(\pi r^2) = 6\pi r^2$,

Cost for the side: $7(2\pi rh) = 14\pi rh$.

So the total cost is $11\pi r^2 + 14\pi rh$ (dollars).

19. From Exercise 7, perimeter = 8x (for rectangle of Exercise 1), hence
$$8x = 40,$$
$$x = 5.$$

Also from Exercise 7 the area is $3x^2$ (for the rectangle of Exercise 1). Therefore, since $x = 5$,
$$\text{Area} = 3(5)^2 = 75 \text{ cm}^2.$$

25. a. $P(x) = R(x) - C(x)$
$$= 21x - (9x + 800) = (12x - 800) \text{ dollars.}$$

b. $x =$ number of sales,
here $x = 120$, so $P(x) = P(120) = 12(120) - 800$
$$= 640 \text{ dollars in profit.}$$

c. Weekly profit function is $P(x) = 12x - 800$ (from (a)),
and we are given a weekly profit of $1000, so

$$12x - 800 = 1000,$$
$$12x = 1800,$$
$$x = 150.$$

Weekly revenue is $R(x) = 21x$,
$$\Rightarrow \text{revenue} = 21(150) = \$3150.$$

31. $(3,162)$, $(6,270)$ are points on the graph $y = f(r)$.
The cost of constructing a cylinder of radius 3 inches is 162 cents, and similarly, for a radius of 6 inches it is 270 cents. Thus the additional cost of increasing the radius from 3 inches to 6 inches is $270 - 162 = 108$ cents.

37. $C(1000) = 4000$.

43. Use Table 1.
Find the x-coordinates of the points on the graph whose y-coordinate is 30,000.

49. $h(t) =$ height of the ball (in feet) after t seconds.
Given that the height $= h(t) = 100$ feet, we want to find the value of the time t. (Example 7 gives an example of a height function.)

So our task is to find the t-coordinates of the points on the graph whose y-coordinate is 100.

Chapter 0: Supplementary Exercises

Study the chapter checklist. Do you understand all the terms in the list? Can you handle the algebra of functions as in Section 0.3? Can you factor quadratic (and simple cubic) polynomials? Have you memorized the quadratic formula? Have you thoroughly (and successfully) practiced using the laws of exponents? If you cannot answer "yes" to all these questions by the time you finish Chapter 1, you are not seriously interested in succeeding in this calculus course, or you need to take a refresher course in college algebra before you study calculus.

The supplementary exercises provide a brief review of the main skills of the chapter.

1. If $f(x) = x^3 + \frac{1}{x}$, then

$$f(1) = 1^3 + \frac{1}{1} = 1 + 1 = 2,$$

$$f(3) = 3^3 + \frac{1}{3} = 27 + \frac{1}{3} = 27\frac{1}{3},$$

$$f(-1) = (-1)^3 + \frac{1}{-1} = -1 - 1 = -2,$$

$$f\left(-\frac{1}{2}\right) = \left(-\frac{1}{2}\right)^3 + \frac{1}{-\frac{1}{2}} = -\frac{1}{8} - 2 = -2\frac{1}{8},$$

$$f(\sqrt{2}) = (\sqrt{2})^3 + \frac{1}{\sqrt{2}} = 2\sqrt{2} + \frac{1}{\sqrt{2}} = 2\sqrt{2} + \frac{\sqrt{2}}{2} = \frac{5\sqrt{2}}{2}.$$

7. The domain of $f(x) = \sqrt{x^2 + 1}$ consists of all values of x since $x^2 + 1$ is never negative.

13. To factor $18 + 3x - x^2$ we first factor out the coefficient -1 of x^2 to get $-1(x^2 - 3x - 18)$. Now, to factor $x^2 - 3x - 18$ we must find c and d such that $cd = -18$ and $c + d = -3$. The solution is $c = -6$ and $d = 3$ and

$$x^2 - 3x - 18 = (x-6)(x+3).$$

Thus

$$18 + 3x - x^2 = -1(x^2-3x-18) = -1(x-6)(x+3).$$

19. $f(x) + g(x) = (x^2 - 2x) + (3x - 1) = x^2 + x - 1.$

25. $f(x) - g(x) = \dfrac{x}{(x^2 - 1)} - \dfrac{(1 - x)}{(1 + x)}$. In order to add two fractions, their denominators must be the same. Observe that $x^2 - 1 = (x+1)(x-1)$. Thus a common denominator for $\dfrac{x}{x^2 - 1}$ and $\dfrac{1 - x}{1 + x}$ is $x^2 - 1$. If we multiply $\dfrac{1 - x}{1 + x}$ by $\dfrac{x - 1}{x - 1}$, we get an equivalent expression whose denominator is $x^2 - 1$. Thus

$$\frac{x}{x^2 - 1} - \frac{1 - x}{1 + x} = \frac{x}{x^2 - 1} - \frac{1 - x}{1 + x} \cdot \frac{x - 1}{x - 1}$$

$$= \frac{x}{x^2 - 1} - \frac{(1 - x)(x - 1)}{x^2 - 1}$$

$$= \frac{x - (-x^2 + 2x - 1)}{x^2 - 1}$$

$$= \frac{x^2 - x + 1}{x^2 - 1}.$$

31. If we substitute g(x) for each occurrence of x in f(x) we obtain f(g(x)). Thus

$$f(g(x)) = \left(\frac{1}{x^2}\right)^2 - 2\left(\frac{1}{x^2}\right) + 4 = \frac{1}{x^4} - \frac{2}{x^2} + 4.$$

37. $(81)^{3/4} = (81^{1/4})^3 = 3^3 = 27,$ (Law 4′)

    $8^{5/3} = (8^{1/3})^5 = 2^5 = 32,$ (Law 4′)

    $(.25)^{-1} = \left(\frac{1}{4}\right)^{-1} = 4.$

43. $\dfrac{x^{3/2}}{\sqrt{x}} = \dfrac{x^{3/2}}{x^{1/2}} = x^{(3/2-1/2)} = x^1 = x.$ (Law 3)

CHAPTER 1

THE DERIVATIVE

1.1 The Slope of a Straight Line

Slope Property 1 will help you understand the concept of the slope of a line. Property 2 is needed when you have to find the slope of a line between two points. Property 3 is the most useful. You must memorize the *point-slope* form of the equation of a line. Exercises 11-16 are simple but very important. Check with your instructor to see how much attention you should give to Slope Properties 4 and 5. They are seldom needed later in the text.

1. If we put the equation

$$y = 2 - 5x,$$

in the form $y = mx + b$, we have

$$y = -5x + 2.$$

Hence the slope is -5.

7. To write the equation

$$2x + 3y = 6,$$

in the form $y = mx + b$, solve for y:

$$3y = -2x + 6,$$

$$y = -\frac{2}{3}x + 2.$$

The slope is $-\frac{2}{3}$.

13. Let $(x_1, y_1) = (5, 0)$ and $m = -7$, and use Slope Property 3.
The equation of the line is

$$y - 0 = -7(x - 5),$$

or

$$y = 35 - 7x.$$

19. By Slope Property 2, the slope of the line is

$$\frac{-2 - 0}{1 - 0} = \frac{-2}{1} = -2.$$

Since $(0,0)$ is on the line, we can use Slope Property 3 to get the equation of the line:

$$y - 0 = -2(x - 0),$$

or

$$y = -2x.$$

25. (a) is matched with (C). To see this we look at the intercepts. The x-intercept is the value of x when $y = 0$. Thus for $x + y = 1$, we see that:

$$y = 0 \Rightarrow x = 1.$$

The y-intercept is the value of y when $x = 0$, so again

$$x = 0 \Rightarrow y = 1.$$

Graph (C) is the only graph which has both intercepts positive.

(b) is matched with (B). Repeating the above process we find that the x-intercept is $x = 1$, and the y-intercept is $y = -1$. The only graph which has a positive x-intercept and a negative y-intercept is (B). By similar reasoning:

(c) is matched with (D).

(d) is matched with (A).

31. Since the slope of this line is 2, if we start at a point on the line and move 1 unit to the right and then 2 units up (in the positive y-direction) we will reach another point on the line. If we start at $(1,3)$ and move 1 unit to the right and 2 units up, we arrive at $(2,5)$. So $(2,5)$ is on the line. A similar move from $(2,5)$ takes us to $(3,7)$, so $(3,7)$ is on the line.

Finally, suppose that $(0,y)$ is on the line. This point is one unit to the left of $(1,3)$ in the x-direction. Starting at $(0,y)$ and moving one unit to the right and 2 units up, we arrive at $(1,y+2)$. This point is on the line and so is $(1,3)$. Hence we must have $y + 2 = 3$, and $y = 1$. Thus $(0,1)$ is on the line.

37. By Slope Property 3,

$$y - (-1) = -2(x - 0)$$

or,
$$y = -2x - 1.$$

To graph this line, first plot the y-intercept $(0,-1)$. Then since the slope is -2, we start at $(0,-1)$ and move one unit right, and then two units in the negative y-direction. The new point $(1,-3)$ is on the line. Draw the straight line through these two points.

43. Parallel lines by definition always have the same slope. So we need to find the slope of $4x + 5y = 6$. Thus
$$5y = -4x + 6,$$
$$y = \frac{-4}{5}x + \frac{6}{5}.$$

Now since $y = mx + b$ is the equation of a line we have that the slope m is $-\frac{4}{5}$.

Using Slope Property 3, we see that the equation of a line through a general point (x_1, y_1) with slope $-\frac{4}{5}$ is, using Slope Property 3:
$$y - y_1 = -\frac{4}{5}(x - x_1),$$
$$y - y_1 = -\frac{4}{5}x + \frac{4}{5}x_1.$$

So,
$$\frac{4}{5}x + y = y + \frac{4}{5}x,$$
$$4x + 5y = 5y_1 + 4x_1.$$

Since $5y_1 + 4x_1$ is any constant C, the possible lines are $4x + 5y = C$.

49. At a point (x,y) the tangent line to the parabola has slope $2x$. The point $\left(-\frac{1}{2}, \frac{1}{4}\right)$ corresponds to $x = -\frac{1}{2}$, so the slope of the tangent line at this point is $2\left(-\frac{1}{2}\right) = -1$. Now, let $(x_1, y_1) = \left(-\frac{1}{2}, \frac{1}{4}\right)$ and $m = -1$, and use Slope Property 3. The equation of the tangent line through $\left(-\frac{1}{2}, \frac{1}{4}\right)$ is
$$y - \frac{1}{4} = -1\left(x - \left(-\frac{1}{2}\right)\right),$$
or,

$$y - \frac{1}{4} = -\left(x + \frac{1}{2}\right).$$

55. By Slope Property 1: m_1 is positive; m_2 is negative.

Slope of ℓ_1 is $\frac{m_1}{1}$; slope of ℓ_2 is $\frac{m_2}{1}$.

We have two right-angled triangles. The first has sides of length a, 1, and m, which by Pythagorus produces

$$a^2 = m_1{}^2 + 1^2 = m_1{}^2 + 1 \quad (a = \text{hypotenuse}).$$

The other right-angled triangle has sides of length b, 1, and $(-m_2)$. (Note: $-m_2$, since m_2 is negative and we can't have a negative length, we make it positive.) So similarly: $b^2 = m_1{}^2 + 1$. Combining these results we obtain:

$$a^2 + b^2 = m_1{}^2 + m_2{}^2 + 2 \quad (*).$$

Since $\ell_1 \perp \ell_2$, we have yet another right-angled triangle with sides of length:

$$a, \ b, \ \text{and} \ (m_1 + (-m_2)),$$
$$\text{i.e., } a, \ b, \ \text{and} \ m_1 - m_2 \ (\text{hypotenuse}).$$

Hence, as before

$$a^2 + b^2 = (m_1 - m_2)^2 = m_1{}^2 + m_2{}^2 - 2m_1 m_2.$$

From (*) we know that $m_1{}^2 + m_2{}^2 + 2 = m_1{}^2 + m_2{}^2 - 2m_1 m_2$. This leads to

$$2 = -2m_1 m_2,$$
$$m_1 m_2 = -1.$$

1.2 The Slope of a Curve at a Point

This brief section should be read carefully. Example 2 and Exercises 23, 24, and 33 are very important, and so we have included a solution of Exercise 23. Of the students who do *not* use this study guide, at least 20% will miss an exam problem on the equation of a tangent line. If you carefully study the solution of Exercise 23 here and the solution of Exercise 43 in Section 1.6, you should have no difficulty on an exam.

1. See the sketch in the answer section of the text.

7. The slope of the curve at the point P is, by definition, the slope of the tangent line at P. If we move one unit in the positive x-direction from P, we can return to the line by moving one unit in the positive y-direction. Therefore the slope is 1.

13. Small positive slope.

19. The slope of the graph of $y = x^2$ at the point (x,y) is 2x. The point $(-2,4)$ corresponds to $x = -2$, so the slope at this point is $2x = 2(-2) = -4$. Now, let $(x_1,y_1) = (-2,4)$ and $m = -4$ and use Slope Property 3. The equation of the tangent line through $(-2,4)$ is:

$$y - 4 = -4(x - (-2)),$$

or,

$$y - 4 = -4(x + 2).$$

23. We will use the point-slope equation of the tangent line to the graph of $y = x^2$ at $x = 1.5$. We need a point (x_1,y_1) and a slope m. There are two basic principles to keep in mind.

(i) Use the *original equation* $y = x^2$ to find a *point* on the graph. When $x = 1.5$, we have $y = (1.5)^2 = 2.25$. Thus $(1.5,2.25)$ is on the graph of $y = x^2$.

(ii) Use the *slope formula* 2x to find the *slope* of the graph at a point. When $x = 1.5$, the slope of the graph is $2(1.5) = 3$. By definition, this slope is the slope of the tangent line to the graph at the point where $x = 1.5$.

The desired tangent line equation has the form $y - y_1 = m(x - x_1)$, where $x_1 = 1.5$, $y_1 = 2.25$, and $m = 3$. That is,

$$y - 2.25 = 3(x - 1.5).$$

Leave the answer in this form unless you are specifically asked to rewrite the answer in some equivalent form such as the slope intercept form.

Warning: A common mistake in Exercise 23 is to replace m in the equation $y - y_1 = m(x - x_1)$ by the slope formula 2x instead of a specific *value* of m. But the equation $y - 2.25 = 2x(x - 1.5)$ is *not* the equation of a line. (In fact, this equation simplifies to $y - 2.25 = 2x^2 - 3x$, or $y = 2x^2 - 3x + 2.25$, which is a quadratic equation.) The equation of a tangent *line* must involve a specific *number* m that gives the slope of the desired tangent line.

25. The slope of the graph of $y = x^2$ at the point (x,y) is $2x$. Hence, if the slope is $\frac{5}{3}$, then

$$2x = \frac{5}{3},$$

$$x = \frac{5}{3} \cdot \frac{1}{2} = \frac{5}{6}.$$

To find the y-coordinate of the point where $x = \frac{5}{6}$, use the original equation $y = x^2$.

$$y = \left(\frac{5}{6}\right)^2 = \frac{25}{36}.$$

Hence, $\left(\frac{5}{6}, \frac{25}{36}\right)$ is the desired point on the graph.

31. The slope formula for the curve $y = x^3$ is given by the formula $3x^2$. The point $\left(-\frac{1}{2}, -\frac{1}{8}\right)$ corresponds to $x = -\frac{1}{2}$, so the slope at $\left(-\frac{1}{2}, -\frac{1}{8}\right)$ is $3x^2 = 3\left(-\frac{1}{2}\right)^2 = 3\left(\frac{1}{4}\right) = \frac{3}{4}$.

36. See the solutions in the text for a drawing.

1.3 The Derivative

Your short-term goal for this section should be to learn the derivative formulas that appear in boxes and to become familiar with the notation introduced on page 96. This will enable you to do the homework for this section. Your long-term goal should be to have some understanding of the secant-line calculation of the derivative, as described on pages 96—98. This material is not grasped easily. Reading it *out loud* will help you to go over the ideas slowly and carefully. Plan to review this section before your first exam and again when you reach Chapter 3.

1. If $f(x) = mx + b$, then $f'(x) = m$. Hence if $f(x) = 2x - 5$, then $f'(x) = 2$.

7. Since $\sqrt[3]{x} = x^{1/3}$, we may apply the power rule with $r = \frac{1}{3}$:

$$f(x) = x^{1/3},$$
$$f'(x) = \frac{1}{3}x^{(1/3)-1} = \frac{1}{3}x^{-2/3}.$$

13. The derivative of a constant function $f(x) = b$ is the zero function. Hence if $f(x) = \frac{3}{4}$, then $f'(x) = 0$.

19. If $f(x) = \frac{1}{x}$, then $f'(x) = -\frac{1}{x^2}$ $(x \neq 0)$, by formula (5) on page 95. Hence at $x = 3$, $f'(3) = -\frac{1}{3^2} = -\frac{1}{9}$.

Helpful Hint: Formula (5) for the derivative of $\frac{1}{x}$ is used so frequently that you should memorize it even though it is only a special case of the power rule.

25. Use the power rule with $r = 4$:

$$\frac{d}{dx}(x^4) = 4x^{4-1} = 4x^3.$$

Setting $x = 3$ in this derivative, we find that the slope of the curve $y = x^4$ at $x = 3$ is $4(3)^3 = 4 \cdot 27 = 108$.

31. If $f(x) = \frac{1}{x^5}$, then $f(2) = \frac{1}{2^5} = \frac{1}{32}$. To compute $f'(2)$, we first determine $f'(x)$ and *then* substitute 2 for x in the expression for $f'(x)$.

Since $\frac{1}{x^5} = x^{-5}$, we may apply the power rule with $r = -5$.

$$f(x) = x^{-5},$$
$$f'(x) = -5x^{-5-1} = -5x^{-6} = -\frac{5}{x^6}.$$
$$f'(2) = -\frac{5}{2^6} = -\frac{5}{64}.$$

37. Using the power rule with $r = 8$, we find

$$\frac{d}{dx}(x^8) = 8x^{8-1} = 8x^7.$$

43. Apply the power rule with $r = \frac{1}{5}$:

$$y = x^{1/5},$$

$$\frac{dy}{dx} = \frac{1}{5}x^{(1/5)-1} = \frac{1}{5}x^{-4/5}.$$

49. The three lines and $y = f(x)$ intersect at the point $(a, f(a))$. Take the line $y = 2.01x - .51$ at this point. Then:

$$f(a) = 2.01a - .51,$$

similarly using $y = 2.02x - .52$,

$$f(a) = 2.02a - .52.$$

So,

$$2.01a - .51 = 2.02a - .52,$$
$$0.01 = 0.01\,a,$$
$$a = 1,$$
$$f(a) = 2.02a - .52 = 2.02 - .52 = 1.5.$$

To estimate $f'(a)$ use the secant line calculation. The secant line $y = 2.01x - .51$ is "nearly" a tangent line, therefore the slope of this line is nearly $f'(a)$.

The slope of the tangent line is slightly less than the secant lines slope of 2.01. Hence,

$$f'(a) \approx 2.$$

55. $f'(x) = \lim_{h \to 0} \dfrac{f(x + h) - f(x)}{h},$ $(h \neq 0)$

Taking $f(x) = x^3$, and substituting into the above equation we have:

$$f'(x) = \lim_{h \to 0} \frac{(x + h)^3 - (x)^3}{h}$$

$$= \lim_{h \to 0} \frac{x^3 + 3x^2h + 3xh^2 + h^3 - x^3}{h}$$

$$= \lim_{h \to 0} 3x^2 + 3xh + h^2 = 3x^2.$$

1.4 Limits and the Derivative

Instructors differ widely about how much theoretical material to include in a course using our text. We recommend that,
at a minimum, you read pages 103—104, even if no exercises
are assigned. The skills practiced in this section are not
needed for later work, but the exercises will give you
valuable experience with simple limits and will improve your
understanding of derivatives.

1. $\lim\limits_{x \to 3} g(x)$ does not exist. As x approaches 3 from the
    right, g(x) approaches 2, but as x approaches 3 from the
    left, the values for g(x) do not approach 2. In order for
    a limit to exist, the values of the function must approach
    the same number as x approaches 2 from each direction.

7. Since 1 - 6x is a polynomial, the limit exists and

$$\lim_{x \to 1} (1 - 6x) = 1 - 6(1) = 1 - 6 = -5.$$

13. Using the Limit Theorems, we have

$$\lim_{x \to 7} (x + \sqrt{x - 6})(x^2 - 2x + 1)$$

$$= \left[\lim_{x \to 7} (x + \sqrt{x - 6})\right]\left[\lim_{x \to 7} (x^2 - 2x + 1)\right] \qquad \text{(Thm. V)}$$

$$= \left[\lim_{x \to 7} x + \lim_{x \to 7} (x - 6)^{1/2}\right]\left[\lim_{x \to 7} (x^2 - 2x + 1)\right] \qquad \text{(Thm. III)}$$

$$= \left[\lim_{x \to 7} x + \left(\lim_{x \to 7} (x - 6)\right)^{1/2}\right]\left[\lim_{x \to 7}(x^2 - 2x + 1)\right] \qquad \text{(Thm. II)}$$

$$= [7 + (1)^{1/2}][49 - 14 + 1] \qquad \begin{bmatrix}\text{Limits of poly-} \\ \text{nomial functions}\end{bmatrix}$$

$$= (8)(36) = 288.$$

19. Since $\dfrac{-2x^2 + 4x}{x - 2} = \dfrac{-2x(x - 2)}{x - 2} = -2x$ for $x \neq 2$,

$$\lim_{x \to 2} \frac{-2x^2 + 4x}{x - 2} = \lim_{x \to 2} -2x = -2(2) = -4.$$

25. No limit exists. Observe that $\lim\limits_{x \to 8} x^2 + 64 = 128$ and
    $\lim\limits_{x \to 8} x - 8 = 0$, as x approaches 8. So the denominator gets

very small and the numerator approaches 128. For example, if x = 8.00001, then the numerator is 128.00016 and the denominator is .00001. The quotient is 12,800,016. As x approaches 8 even more closely, the quotient gets arbitrarily large and cannot possibly approach a limit.

31. Since $f'(a) = \lim_{h \to 0} \dfrac{f(a + h) - f(a)}{h}$, we must calculate

$$\lim_{h \to 0} \frac{f(0 + h) - f(0)}{h}.$$

If $f(x) = x^3 + 3x + 1$, then

$$\frac{f(0+h) - f(0)}{h} = \frac{[(0+h)^3 + 3(0+h) + 1] - [0^3 + 3(0) + 1]}{h}$$

$$= \frac{h^3 + 3h + 1 - 1}{h} = \frac{h^3 + 3h}{h} = h^2 + 3.$$

Therefore, $f'(0) = \lim_{h \to 0} (h^2 + 3) = 0^2 + 3 = 3.$

37. We must calculate $f'(0) = \lim_{h \to 0} \dfrac{f(0 + h) - f(0)}{h}.$

If $f(x) = \sqrt{1 - x^2}$, then

$$\frac{f(0 + h) - f(0)}{h} = \frac{\sqrt{1 - (0 + h)^2} - \sqrt{1 - 0^2}}{h}$$

$$= \frac{\sqrt{1 - h^2} - 1}{h}.$$

In order to find $\lim_{h \to 0} \dfrac{\sqrt{1 - h^2} - 1}{h}$, we must apply an algebraic trick since the denominator of this expression approaches zero. We multiply numerator and denominator by $\sqrt{1 - h^2} + 1$ and use the fact that $(a - 1)(a + 1) = a^2 - 1$, where $a = \sqrt{1 - h^2}$.

$$\frac{\sqrt{1 - h^2} - 1}{h} \cdot \frac{\sqrt{1 - h^2} + 1}{\sqrt{1 - h^2} + 1} = \frac{(1 - h^2) - 1}{h\left(\sqrt{1 - h^2} + 1\right)}$$

$$= \frac{-h^2}{h\left(\sqrt{1 - h^2} + 1\right)}$$

$$= \frac{-h}{\sqrt{1 - h^2} + 1}.$$

Now $f'(0) = \lim_{h \to 0} \dfrac{-h}{\sqrt{1 - h^2} + 1} = \dfrac{-0}{\sqrt{1 - 0^2} + 1} = \dfrac{0}{2} = 0.$

43. We want to find $f(x)$ so that $\lim_{h \to 0} \dfrac{\frac{1}{10 + h} - .1}{h}$ has the same form as

$$f'(a) = \lim_{h \to 0} \frac{f(a + h) - f(a)}{h}.$$

So we want $f(a+h) = \dfrac{1}{10 + h}$; and $f(a) = 0.1 = \dfrac{1}{10}$. From this we see that $f(x) = \dfrac{1}{x}$ and $a = 10$. Note that $f(a + h) = f(10 + h) = 1/(10 + h).$

49. As x increases without bound so does x - 8. Therefore, $\dfrac{1}{x - 8}$ approaches zero as x approaches ∞. That is,

$$\lim_{x \to \infty} \frac{1}{x - 8} = 0.$$

1.5 Differentiability and Continuity

We want you to be aware that real applications sometimes involve functions that may not be differentiable at one or more points in their domains. So this section gives you a rare opportunity to see functions whose graphs are not as "nice" as the ones we usually consider.

1. No, the graph in Fig. 6 is not continuous at x = 0 since the limit as x approaches zero does not exist.

7. No, the graph in Fig. 6 is not differentiable at x = 0 because the function is not even *defined* at x = 0. Even if the function were given some value at x = 0, say f(0) = 1, the graph would have a vertical tangent line at x = 0.

13. Since $f(x) = x^2$, the power rule gives a slope-formula, namely, $2x$, that is valid for all x. This slope-formula was verified in Section 1.3 using the limit of the slopes of secant lines. Thus $f(x)$ has a derivative (in the formal sense of Section 1.4) for all x. In particular, $f(x)$ is differentiable at $x = 1$. By Theorem 1 on page 116, $f(x)$ is necessarily continuous at $x = 1$.

19. At $x = 1$ the function $f(x)$ is defined, namely $f(1) = 0$. When computing $\lim_{x \to 1} f(x)$, we exclude consideration of the value $x = 1$; therefore,

$$\lim_{x \to 1} f(x) = \lim_{x \to 1} \frac{1}{x-1} \quad \text{which does not exist.}$$

Hence $\lim_{x \to 1} f(x) \neq f(1)$, so $f(x)$ is not continuous at $x = 1$.

By Theorem 1, since $f(x)$ is not continuous at $x = 1$, it cannot be differentiable at $x = 1$.

25. $f(x)$ is continuous at $x = a$, if

$$\lim_{x \to a} f(x) = f(a), \quad \text{here } f(x) = \frac{(6 + x)^2 - 36}{x}, \quad x \neq 0.$$

Where $a = 0$, we have:

$$\lim_{x \to 0} \frac{(6 + x)^2 - 36}{x}$$

$$= \lim_{x \to 0} \frac{36 + 12x + x^2 - 36}{x}$$

$$= \lim_{x \to 0} (12 + x) = 12.$$

so define $f(0) = 12$, and this definition will make $f(x)$ continuous for all x.

1.6 Some Rules for Differentiation

The rules in this section are mastered by working lots of problems. Make sure you learn to distinguish between a constant that is *added* to a function and a constant that *multiplies* a function. For instance, $x^3 + 5$ is the sum of the cube function $f(x) = x^3$ and the constant function $g(x) = 5$. Thus

$$\frac{d}{dx}(x^3 + 5) = \frac{d}{dx}(x^3) + \frac{d}{dx} 5 \qquad \text{(Sum rule)}$$

$$= 3x^2 + 0,$$

because the derivative of a constant function is zero. However, the constant 5 in the formula $5x^3$ *multiplies* the function x^3. Hence, by the constant-multiple rule,

$$\frac{d}{dx}5x^3 = 5 \cdot \frac{d}{dx}(x^3) = 5 \cdot 3x^2 = 15x^2.$$

Exercise 43 is very important. But before you look at the solution here, go back and read the solution to Exercise 17 in Section 1.2. Then try Exercise 43 in Section 1.6 by yourself. Peek at the solution only if you get stuck.

1. To differentiate $y = x^3 + x^2$, let $f(x) = x^3$ and $g(x) = x^2$ and apply the sum rule.

$$\frac{dy}{dx} = \frac{d}{dx}(x^3 + x^2) = \frac{d}{dx}(x^3) + \frac{d}{dx}(x^2) = 3x^2 + 2x.$$

7. $\frac{d}{dx}(x^4 + x^3 + x) = \frac{d}{dx}(x^4) + \frac{d}{dx}(x^3 + x)$ (Sum rule)

$$= \frac{d}{dx}(x^4) + \frac{d}{dx}(x^3) + \frac{d}{dx}(x) \qquad \text{(Sum rule)}$$

$$= 4x^3 + 3x^2 + 1.$$

13. Write $\frac{4}{x^2}$ in the form $4 \cdot x^{-2}$. Then

$$\frac{dy}{dx} = \frac{d}{dx}(4 \cdot x^{-2})$$

$$= 4 \cdot \frac{d}{dx}(x^{-2}) \qquad \text{(Constant-multiple rule)}$$

$$= 4(-2)x^{-3} \qquad \text{(Power rule)}$$

$$= -8x^{-3}, \text{ or } -\frac{8}{x^3}.$$

19. Since $-\frac{1}{5x^5} = -\frac{1}{5} \cdot \frac{1}{x^5} = \left(-\frac{1}{5}\right) \cdot x^{-5}$, we use the constant-multiple rule and the power rule.

$$\frac{dy}{dx} = \frac{d}{dx}\left(-\frac{1}{5} \cdot x^{-5}\right) = -\frac{1}{5} \cdot \frac{d}{dx}(x^{-5}) = -\frac{1}{5}(-5)x^{-6} = x^{-6}, \text{ or } \frac{1}{x^6}.$$

Warning: Problems like those in Exercises 13, 19, and 29 tend to be missed on exams by many students. The difficulty lies in the first step—recognizing how to write the function as a constant times a power of x. Be sure to review this before the exam.

25. $\dfrac{d}{dx} \, 5\sqrt{3x^3 + x} = \dfrac{d}{dx} \, 5(3x^3 + x)^{1/2}$

$\qquad\qquad = \dfrac{5}{2}(3x^3 + x)^{-1/2} \cdot \dfrac{d}{dx}(3x^3 + x)$ (Gen power rule)

$\qquad\qquad = \dfrac{5}{2}(3x^3 + x)^{-1/2}\left[\dfrac{d}{dx}(3x^3) + \dfrac{d}{dx}(x)\right]$ (Sum rule)

$\qquad\qquad = \dfrac{5}{2}(3x^3 + x)^{-1/2}\left[3 \cdot \dfrac{d}{dx}(x^3) + \dfrac{d}{dx}(x)\right]$

$\qquad\qquad\qquad\qquad\qquad\qquad\qquad$ (Constant-multiple rule)

$\qquad\qquad = \dfrac{5}{2}(3x^3 + x)^{-1/2}[3(3x^2) + 1] = \dfrac{45x^2 + 5}{2\sqrt{3x^3 + x}}\,.$

Warning: Don't forget to use parentheses (or brackets) when appropriate in the general power rule. The next-to-last line in the solution above is incorrect if written in the form:

$\qquad\qquad = \dfrac{5}{2}(3x^3 + x)^{-1/2} \cdot 3(3x^2) + 1.$

$\qquad\qquad\qquad\qquad\qquad$ missing brackets

31. Note that $y = \dfrac{2}{1 - 5x} = 2 \cdot \dfrac{1}{1 - 5x} = 2 \cdot (1 - 5x)^{-1}$. Hence

$\qquad \dfrac{dy}{dx} = \dfrac{d}{dx}(2 \cdot (1 - 5x)^{-1})$

$\qquad\qquad = 2 \cdot \dfrac{d}{dx}(1 - 5x)^{-1}$ (Constant-multiple rule)

$\qquad\qquad = 2(-1)(1 - 5x)^{-2} \cdot \dfrac{d}{dx}(1 - 5x)$ (General power rule)

$\qquad\qquad = -2(1 - 5x)^{-2}(-5) = 10(1 - 5x)^{-2}.$

Warning: Forgetting the −5 in the next-to-the-last line above is a common mistake. Another common error is to carelessly forget the parentheses and write the derivative as

$$-2(1 - 5x)^{-2} - 5.$$

This is definitely incorrect, and many instructors will give no part credit for such an answer. Your ability to find a derivative is of no value to anyone if your algebra is careless and incorrect.

In this *Study Guide*, we'll point where algebra errors are likely to occur so you can guard against them. One key to avoiding such errors on exams is to practice working carefully on your homework.

37. If $f(x) = \left(\dfrac{\sqrt{x}}{2} + 1\right)^{3/2}$ then

$$f'(x) = \frac{d}{dx}\left(\frac{\sqrt{x}}{2} + 1\right)^{3/2}$$

$$= \frac{3}{2}\left(\frac{\sqrt{x}}{2} + 1\right)^{1/2} \cdot \frac{d}{dx}\left(\frac{\sqrt{x}}{2} + 1\right) \qquad \text{(General power rule)}$$

$$= \frac{3}{2}\left(\frac{\sqrt{x}}{2} + 1\right)^{1/2}\left[\frac{d}{dx}\left(\frac{\sqrt{x}}{2}\right) + \frac{d}{dx}(1)\right] \qquad \text{(Sum rule)}$$

$$= \frac{3}{2}\left(\frac{\sqrt{x}}{2} + 1\right)^{1/2}\left[\frac{1}{2} \cdot \frac{d}{dx}(x^{1/2}) + \frac{d}{dx}(1)\right]$$
$$\text{(Constant-multiple rule)}$$

$$= \frac{3}{2}\left(\frac{\sqrt{x}}{2} + 1\right)^{1/2}\left[\frac{1}{2}\left(\frac{1}{2}\right)x^{-1/2} + 0\right]$$

$$= \frac{3}{2}\left(\frac{\sqrt{x}}{2} + 1\right)^{1/2}\left(\frac{1}{4}x^{-1/2}\right).$$

43. The general slope-formula for the curve $y = (x^2 - 15)^6$ is given by the derivative.

$$\frac{dy}{dx} = 6(x^2 - 15)^5 \cdot \frac{d}{dx}(x^2 - 15) \qquad \text{(General power rule)}$$

$$= 6(x^2 - 15)^5(2x)$$

$$= 12x(x^2 - 15)^5.$$

To find the *slope* m of the tangent line at the particular point where x = 4, substitute 4 for x in the *derivative* formula to get

$$m = 12(4)[(4)^2 - 15]^5 = 48(1)^5 = 48.$$

For the equation of the tangent line at x = 4, use the

point-slope form. We already have the slope m = 48. To find a *point* (x_1, y_1) on the curve when $x_1 = 4$, substitute 4 for x in the *original* equation to get

$$y_1 = [(4)^2 - 15]^6 = (1)^6 = 1.$$

The equation of the tangent line is

$$y - 1 = 48(x - 4).$$

Warning: Exercise 43 helped you to get the equation of the tangent line by asking you first for the slope of the line. Exam questions tend to be more like Exercise 44. You are expected to know that you need to find the slope of the line and a point on the line. [Try Exercise 44; the answer is $y - 1 = -\frac{5}{8}(x - 2)$.] Also, see the solution to Exercise 17 in Section 1.2.

49. $f(5) = 2$, $g(5) = 4$, and $f'(5) = 3$, $g'(5) = 1$. We are given that,

$$h(x) = 3f(x) + 2g(x),$$
$$h(5) = 3f(5) + 2g(5) = 3(2) + 2(4) = 14,$$
$$h'(x) = 3f'(x) + 2g'(x),$$
$$h'(5) = 3f'(5) + 2g'(5) = 3(3) + 2(1) = 11.$$

55. The point of intersection of the tangent line and f(x) is (4,5), which alternatively can be stated as f(4) = 5. The tangent line goes through (0,3) and (4,5), which means its slope is $m = \frac{5 - 3}{4 - 0} = \frac{2}{4} = \frac{1}{2}$, or

$$f'(4) = \frac{1}{2}.$$

1.7 More About Derivatives

Your ability to differentiate functions easily and correctly depends upon how much you practice. Use the exercises in this section to gain the experience you need. Don't be annoyed that you have to learn so much notation. We want you to be able to read technical articles in your field and be familiar with some of the notation you will find there.

1. $f(t) = (t^2 + 1)^5$

$$f'(t) = 5(t^2 + 1)^4 \cdot \frac{d}{dt}(t^2 + 1)$$

$$= 5(t^2 + 1)^4 2t = 10t(t^2 + 1)^4.$$

7. $\frac{d}{dP}(3P^2 - \frac{1}{2}P + 1) = \frac{d}{dP}(3P^2) + \frac{d}{dP}\left(-\frac{1}{2}P\right) + \frac{d}{dP}(1)$

$$= 3\cdot\frac{d}{dP}(P^2) + -\frac{1}{2}\cdot\frac{d}{dP}(P) + \frac{d}{dP}(1)$$

$$= 3(2P) - \frac{1}{2}(1) + 0$$

$$= 6P - \frac{1}{2}.$$

Helpful Hint: The notation $\frac{d}{dt}$ in Exercise 9 indicates that t is the independent variable when you differentiate. Any other letters appearing in the function represent constants, even though the values of these constants are not specified.

13. $y = \sqrt{x} = x^{1/2}$,

$\frac{dy}{dx} = \frac{1}{2}x^{-1/2}$,

$\frac{d^2y}{dx^2} = \frac{d}{dx}\left(\frac{dy}{dx}\right) = \frac{d}{dx}\left(\frac{1}{2}x^{-1/2}\right) = \frac{1}{2}\left(-\frac{1}{2}\right)x^{-3/2} = -\frac{1}{4}x^{-3/2}.$

Helpful Hint: Problems involving radical signs (mainly square roots) are handled more easily when you switch to exponential notation, as in the solution to Exercise 13, above. Your instructor will probably permit you to write your answers in this form, too.

19. $f(P) = (3P + 1)^5$,

$f'(P) = 5(3P + 1)^4\cdot\frac{d}{dP}(3P + 1) = 5(3P + 1)^4\cdot 3$

$$= 15(3P + 1)^4. \quad \text{[Don't forget this.]}$$

$f''(P) = \frac{d}{dP}f'(P) = \frac{d}{dP}\left[15(3P + 1)^4\right] = 15\cdot\frac{d}{dP}(3P + 1)^4$

$$= 15\cdot 4(3P + 1)^3\cdot\frac{d}{dP}(3P + 1) = 60(3P + 1)^3(3)$$

$$= 180(3P + 1)^3.$$

25. $\frac{d}{dx}(3x^3 - x^2 + 7x - 1)$

$$= \frac{d}{dx}(3x^3) + \frac{d}{dx}(-x^2) + \frac{d}{dx}(7x) + \frac{d}{dx}(-1)$$

$$= 9x^2 - 2x + 7 + 0$$

$$= 9x^2 - 2x + 7.$$

$$\frac{d^2}{dx^2}(3x^2 - x^2 + 7x - 1) = \frac{d}{dx}(9x^2 - 2x + 7)$$

$$= 18x - 2 + 0$$

$$= 18x - 2.$$

$$\left.\frac{d^2}{dx^2}(3x^2 - x^2 + 7x - 1)\right|_{x=2} = 18(2) - 2 = 36 - 2 = 34.$$

31. $\frac{dR}{dx} = \frac{d}{dx}(1000 + 80x - .02x^2)$

$$= 0 + 80 - (.02)(2x)$$

$$= 80 - .04x.$$

$$\left.\frac{dR}{dx}\right|_{x=1500} = 80 - .04(1500) = 80 - 60 = 20.$$

1.8 The Derivative as a Rate of Change

This section is crucial for later work. There are three main categories of problems: (1) the rate of change of some function—either an abstract function or a function in an application where some quantity is changing with respect to time; (2) the rate of change of one economic quantity with respect to another—problems involving the adjective "marginal"; and (3) velocity and acceleration problems.

Problems in one category may seem quite different from those in another category. But this difference is superficial and is due to the terminology involved. Try to discover similarities in the problems. The key to this lies in the Practice Problems. Remember to use the practice problems correctly. Try to answer all four problems yourself before you look at the solutions.

1. (a) The average rate of change = $\frac{\Delta y}{\Delta x}$. On the interval 0 to 2 we see that:

$$f(0) = 0, \quad f(2) = 16,$$

which in turn gives us the points $(0,0)$ and $(2,16)$, so

$$\frac{\Delta y}{\Delta x} = \frac{16 - 0}{2 - 0} = 8.$$

On the interval 0 to 1:
$$f(0) = 0; \quad f(1) = 4.$$
So now the points are $(0,0)$ and $(1,4)$, hence

$$\frac{\Delta y}{\Delta x} = \frac{4 - 0}{1 - 0} = 4.$$

Finally on the interval 0 to 0.5:
$$f(0) = 0; \quad f(0.5) = 4(0.5)^2 = 1,$$
giving the points $(0,0)$ and $(0.5,1)$, so

$$\frac{\Delta y}{\Delta x} = \frac{1 - 0}{0.5} = 2.$$

(b) The instantaneous rate of change of $f(x)$ at $x = 0$ is $f'(0)$. Since $f'(x) = 8x \Rightarrow f'(0) = 0$.

7. (a) $W(t) = .1t^2$, therefore $W'(t) = .2t$. This means that the instantaneous rate of growth when $t = 5$ is $W'(5)$;

$$\text{i.e., } W'(5) = .2(5) = 1 \text{ gram/week.}$$

(b) $W'(t) = 5$, but $W'(t) = .2t$, so $.2t = 5$, hence
$$t = 25 \text{ weeks.}$$

13. $f(a + \Delta x) - f(a) \approx f'(a) \cdot \Delta x$. Rearranging produces

$$f(a + \Delta x) \approx f(a) + f'(a) \cdot \Delta x. \qquad (*)$$

Here $f'(a)$ is negative and small; Δx is positive and small, therefore $f'(a) \cdot \Delta x$ is negative and also small. Thus $f(a + \Delta x)$ by (*) is slightly smaller than $f(a)$ which means the statement is true.

19. $f(3) = 170$ is the temperature 3 minutes after the cup of coffee is poured.
$f'(3) = -5$ is the instantaneous rate of change of temperature 3 minutes after the cup of coffee was poured.

25. (a) $C(x) = .1x^3 - 6x^2 + 136x + 200,$
$$C(21) = .1(21)^3 - 6(21)^2 + 136(21) + 200 = 1336.1,$$

$$C(20) = .1(20)^3 - 6(20)^2 + 136(20) + 200 = 1320.0,$$

$$C(21) - C(20) = 1336.1 - 1320.0 = 16.1 \text{ dollars}.$$

This is the extra cost of raising the production from 20 to 21 units.

(b) The true marginal cost involves the derivative:

$$C'(x) = .3x^2 - 12x + 136,$$
$$C'(20) = .3(20)^2 - 12(20) + 136$$
$$= 16.0 \text{ dollars per unit}.$$

31. This exercise, along with #33, is a key problem. It is essential that you try all five parts without looking at the text or the solutions below. If you must have help, look again at the practice problems. After you have tried Exercise 31, compare your answers with those below. Also look again at the practice problems.

(a) Since velocity is the rate of change in position, the initial velocity of the rocket when t = 0 is s'(0).

$$s(t) = 160t - 16t^2,$$

$$s'(t) = 160 - 32t,$$

$$s'(0) = 160 - 32(0) = 160.$$

So the velocity at t = 6 is 10 km/hr.

(b) "Velocity after 2 seconds" means velocity when t = 2.

$$v(2) = s'(2) = 160 - 32(2) = 96 \text{ ft/sec}.$$

(c) Acceleration involves the rate of change of velocity, that is, the derivative of v(t) = 160 - 32t.

$$a(t) = v'(t) = -32.$$

Since the distance is in feet and time is in seconds, the acceleration is -32 feet per second per second. If the units were kilometers and hours, as in Exercise 23, the acceleration would be measured in kilometers per hour per hour.

(d) In this part the time is unknown. The rocket hits the ground when the distance above the ground is zero. So set s'(t) = 0 and solve for t:

$$160t - 16t^2 = 0,$$

$$16t(10 - t) = 0,$$

$$t = 0, \text{ and } t = 10.$$

The rocket slams into the ground when t = 10 seconds.

(e) We know the time from (d). The velocity at t = 10 seconds is

$$v(t) = 160 - 32(10) = -160 \text{ ft/sec.}$$

A negative sign on the velocity indicates that the distance function is decreasing, that is, the rocket is falling down.

Helpful Hint: Exercise 31(e) is a good exam question. If 31(d) is not on the exam, too, here is how to analyze the problem:

1. "At what velocity" means you must find some value of the velocity. To do this you need the velocity function and some specific time t.

2. You can get the velocity function by computing s'(t).

3. Since you aren't given the time, you must determine the time from the fact that the rocket has just smashed into the ground. That is, you must solve a question like #31(d), even though it may not be listed specifically on the exam.

Helpful Hint: Check with your instructor about whether your test answers for velocity and acceleration problems must include the correct units.

37. (a) f(6) = 1790.85 is the balance after 10 years with an interest rate of 6%.

f'(6) = 168.95 gives the instantaneous rate of change of the balance when r = 6% is $168.95.

(b) We can approximate the balance when r = 6.2 by using the formula: $f(a + \Delta x) \approx f(a) + f'(a) \cdot \Delta x$.
Here a = 6, Δx = .2, so

$$f(6.2) = f(6 + 0.2) \approx f(6) + f'(6) \cdot (0.2)$$
$$= 1790.85 + 168.95(0.2) = \$1824.64$$

(c) Now a = 6, Δx = -0.3, and similarly to (b), we see that a + Δx = 5.7, so
$$f(5.7) \approx f(6) + f'(6)(-.03)$$
$$= 1790.85 + 168.95(-0.3)$$
$$= 1790.85 - 50.685 = \$1740.17.$$

Review of Chapter 1

The derivative is presented in three different ways. The derivative is defined *geometrically* in Section 1.3.

> The derivative of f(x) is a function f'(x) whose value at x = a gives the slope of the graph of f(x) at x = a.

The secant-line approximation describes how derivative formulas are obtained. This is made more precise in Section 1.4 when the derivative is defined *analytically* as the limit of difference quotients.

$$f'(a) = \lim_{h \to 0} \frac{f(a + h) - f(a)}{h}$$

Finally, in Section 1.8 the derivative is described *operationally* as the rate of change of a function. Your review of Chapter 1 should include an attempt to see how these three aspects of the derivative concept are related. At the same time, of course, you should review the basic techniques for *calculating* derivatives (Sections 1.6 and 1.7). The supplementary exercises will help here. Some instructors tend to look at them when they prepare exams. Finally, don't forget to review Section 1.1.

Chapter 1: Supplementary Exercises

1. Use the point-slope equation $y - y_1 = m(x - x_1)$, with $(x_1, y_1) = (0,3)$ and $m = -2$. That is,

$$y - 3 = -2(x - 0),$$
   or
$$y = -2x + 3.$$

   To graph this line, first plot the y-intercept (0,3). Then move one unit to the right and then two units in the negative y-direction. The new point (1,1) will also be on the line. Draw the straight line through these two points. (See sketch in the answer section of the text.)

7. Apply Slope Property 2 with $(x_1, y_1) = (-1, 4)$ and $(x_2, y_2) = (3, 7)$. The slope of the line is

$$\frac{y_2 - y_1}{x_2 - x_1} = \frac{7 - 4}{3 - (-1)} = \frac{3}{4}.$$

Now setting $(x_1, y_1) = (-1, 4)$ and $m = \frac{3}{4}$ in the point-slope equation of a line, we have

$$y - 4 = \frac{3}{4}(x - (-1)),$$

or

$$y = \frac{3}{4}x + \frac{19}{4}.$$

To graph this line plot the points $(-1, 4)$ and $(3, 7)$ and then draw the straight line through these two points. (See sketch in the answer section of the text.)

13. The y-axis has equation $x = 0$

y-axis $(x = 0)$

x-axis.

19. $\frac{d}{dx}\left(\frac{3}{x}\right) = \frac{d}{dx}(3 \cdot x^{-1}) = 3 \cdot \frac{d}{dx}(x^{-1})$ (Constant-multiple rule)

$\quad = 3(-1)x^{-2} = -3x^{-2},$ or $-\frac{3}{x^2}.$

25. $\frac{d}{dx}\sqrt{x^2 + 1} = \frac{d}{dx}(x^2 + 1)^{1/2}$

$\quad = \frac{1}{2}(x^2 + 1)^{-1/2} \cdot \frac{d}{dx}(x^2 + 1)$ (Gen. power rule)

$\quad = \frac{1}{2}(x^2 + 1)^{-1/2}(2x) = x(x^2 + 1)^{-1/2}$

$\quad = \frac{x}{\sqrt{x^2 + 1}}.$

31. If $f(x) = [x^5 - (x - 1)^5]^{10},$ then

$f'(x) = 10[x^5 - (x - 1)^5]^9 \cdot \frac{d}{dx}[x^5 - (x - 1)^5]$
$\qquad\qquad\qquad\qquad\qquad\qquad$ (Gen. power rule)

$\quad = 10[x^5 - (x - 1)^5]^9 \left[\frac{d}{dx}(x^5) - \frac{d}{dx}(x - 1)^5\right]$
$\qquad\qquad\qquad\qquad\qquad\qquad\qquad$ (Sum rule)

1-23

$$= 10[x^5 - (x - 1)^5]^9[5x^4 - 5(x - 1)^4 \cdot \frac{d}{dx}(x - 1)]$$

(Gen. power rule)

$$= 10[x^5 - (x - 1)^5]^9[5x^4 - 5(x - 1)^4].$$

37. If $h(x) = \frac{3}{2}x^{3/2} - 6x^{2/3}$, then

$$h'(x) = \frac{3}{2} \cdot \frac{d}{dx}(x^{3/2}) - 6 \cdot \frac{d}{dx}(x^{2/3})$$

$$= \frac{3}{2}\left(\frac{3}{2}\right)x^{1/2} - 6\left(\frac{2}{3}\right)x^{-1/3}$$

$$= \frac{9}{4}x^{1/2} - 4x^{-1/3}.$$

43. If $f(x) = x^{5/2}$, then

$$f'(x) = \frac{5}{2}x^{3/2}$$

and

$$f''(x) = \left(\frac{5}{2}\right)\left(\frac{3}{2}\right)x^{1/2} = \frac{15}{4}\sqrt{x}.$$

Hence $f''(4) = \frac{15}{4}\sqrt{4} = \frac{15}{4}(2) = \frac{15}{2}.$

49. $\frac{d}{dP}(\sqrt{1 - 3P}) = \frac{d}{dP}(1 - 3P)^{1/2}$

$$= \frac{1}{2}(1 - 3P)^{-1/2} \cdot \frac{d}{dP}(1 - 3P) \quad \text{(Gen. power rule)}$$

$$= \frac{1}{2}(1 - 3P)^{-1/2}(-3)$$

$$= -\frac{3}{2}(1 - 3P)^{-1/2}.$$

55. $\frac{d}{dt}(t^3 + 2t^2 - t) = \frac{d}{dt}(t^3) + 2 \cdot \frac{d}{dt}(t^2) - \frac{d}{dt}(t)$

$$= 3t^2 + 4t - 1.$$

$\frac{d^2}{dt^2}(t^3 + 2t - t) = \frac{d}{dt}(3t^2 + 4t - 1) = 6t + 4.$

$\frac{d^2}{dt^2}(t^3 + 2t - t)\Big|_{t=-1} = 6(-1) + 4 = -2.$

61. The slope of the graph of $y = x^2$ at the point (x,y) is $2x$, hence the slope of the tangent line at $\left(\frac{3}{2}, \frac{9}{4}\right)$ is $2\left(\frac{3}{2}\right) = 3$. Let $(x_1, y_1) = \left(\frac{3}{2}, \frac{9}{4}\right)$ and $m = 3$ and use the point-slope equation for the tangent line to get

$$y - \frac{9}{4} = 3\left(x - \frac{3}{2}\right),$$

or

$$y = 3x - \frac{9}{4}.$$

Now sketch the curve $y = x^2$ by plotting a few points, including the point $\left(\frac{3}{2}, \frac{9}{4}\right)$. Move one unit to the right of $\left(\frac{3}{2}, \frac{9}{4}\right)$ and then 3 units in the positive y-direction to reach another point on the tangent line. Draw the straight line through these two points. (See the answer section of the text.)

Helpful Hint: Make sure you can work Exercises 61 and 63 without any help. Make a note to work 63 later when you have not just finished reading this study guide. If you need help, see the solutions of Exercise 23 in Section 1.2 and Exercise 43 in Section 1.6.

67. $s(t) = -16t^2 + 32t + 128,$ (height of binoculars)

$s'(t) = -32t + 32.$ (velocity of binoculars)

To answer the question "How fast ..." we must give the value of the velocity, $s'(t)$. But at what time? The time is identified only by the phrase, "when they hit the ground." Can we describe this time using $s(t)$ or $s'(t)$? Yes, since $s(t) = 0$ when the binoculars are on the ground. So we set $s(t)$ equal to zero and solve to find the time when this happens.

$$-16t^2 + 32t + 128 = 0,$$
$$-16(t^2 - 2t - 8) = 0,$$
$$-16(t - 4)(t + 2) = 0.$$

Thus $t - 4 = 0$ or $t + 2 = 0$, so that $t = 4$ or $t = -2$. We discard the possibility $t = -2$ because the appropriate domain for the function $s(t)$ is $t \geq 0$. The binoculars hit the ground when $t = 4$ seconds, and their velocity at that time is $s'(4) = -32(4) + 32 = -128 + 32 = -96$ feet per second. The negative sign indicates that the distance above the ground is decreasing. So the binoculars are *falling* at the rate of 96 feet per second.

Helpful Hint: Exercise 67 is a typical exam question. The solution require two steps—finding $t = 4$ and computing $s'(4)$. Students who miss this problem usually try to work the problem in one step and don't know where to begin.

73. (a) $f'(x)$ is positive because as x increases the number of gallons used increases; i.e., the rate of change of gallons used is positive.

    (b) $f'(x)$ is greater for the limousine, since a limousine uses more gallons of gas than a subcompact car.

    (c) $f(a + \Delta x) - f(a) \approx f'(a) \cdot \Delta x$.

    Here $a = 25,000$, and $\Delta x = 5$, so

    $$f(25,005) - f(25,000) \approx f'(25,000)5$$
    $$= (.04)5 = 0.2 \text{ gallons.}$$

79. $x^2 - 8x + 16 = (x - 4)^2$, so

$$\lim_{x \to 4} \frac{x - 4}{x^2 - 8x + 16} = \lim_{x \to 4} \frac{x - 4}{(x - 4)^2} = \lim_{x \to 4} \frac{1}{x - 4},$$

which does not exist since the denominator is 0 at $x = 4$ and $\frac{1}{0}$ is not defined.

CHAPTER 2

APPLICATIONS
OF THE
DERIVATIVE

This chapter culminates with one of the most powerful applications of the derivative—the solution of optimization problems. Although graphs play a key role in these problems, graphs are important in their own right. Sections 2.1 and 2.2 provide the background for the key topics covered in the remaining sections of the chapter.

2.1 Describing Graphs of Functions

The terms defined in this section are used throughout the text. The words *maximum* and *minimum* are used to refer to both points and values of a function. When referring to a point, they are usually preceded with the word "relative." When referring to values, "maximum value" means the largest value that the function assumes on its domain, and "minimum value" means the smallest value the function assumes on its domain.

1. Functions a, e, and f are increasing for all x since the graphs rise as we move along them from left to right.

7. Since the graph is superimposed on graph paper, we see that (0,2) is a relative minimum point, (1,3) is an inflection point, and (2,4) is a relative maximum point. Therefore, the key features of the graph are as follows:

    (a) Decreasing for x < 0.
    (b) Relative minimum point at x = 0.
    (c) Increasing for 0 < x < 2.
    (d) Relative maximum point at x = 2.
    (e) Decreasing for x > 2.
    (f) Concave up for x < 1, concave down for x > 1.
    (g) Inflection point at (1,3).
    (h) y-intercept (0,2), x-intercept at about (3.4,0).

13. Take an arbitrary point on the left-hand side of the
    graph, say x = -5, and gradually increase its value. We
    see initially that the slope to the graph is negative.
    As x increases from -5 to x = -2 this slope becomes less
    negative; i.e., it increases until finally at the point
    (-2,-2) the slope is zero. This point is the relative
    minimum. Continuing, as x goes from -2 to x = 2, the
    slope is positive and increases. So in conclusion, the
    graph is concave up for all x ≠ -2, hence there are no
    inflection points.

19.

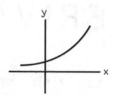

    It is clear that both f(x) and its slope increase as x
    increases, because the curve is rising and it lies above
    the tangent line at each point.

25.

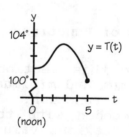

31. The graph is concave down from t = 1920 to t = 1960 and
    concave up from t = 1960 to t = 1990. The term "decreas-
    ing most rapidly" means, graphically, the place of maximum
    negative slope, and this occurs at the inflection point;
    i.e., when t = 1960. Translating, the number of farms was
    decreasing most rapidly when t = 1960.

37.

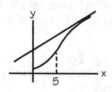

2.2   The First and Second Derivative Rules

The exercises in this section provide a warm-up to the problem of sketching the graphs of functions. Here we gain insights into the types of questions to ask in order to determine the main features of the graph of a given function.

1.   If a function has a positive first derivative for all x, the First Derivative Rule tells us that the function is increasing for all values of x. This is true of graphs b, c, and f.

7.   The only specific point on the graph is (2,1). We plot this point and then use the fact that f'(2) = 0 to sketch the tangent line at x = 2. Since the graph is concave up, the point (2,1) must be a minimum point. Since it is concave up for all x, there are no other relative extreme points or inflection points.

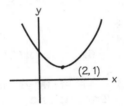

13.   f(3) = 4 ⇒ (3,4) on the graph.
       f'(3) = -1/2, so the slope of the graph at (3,4) is -1/2.
       f"(3) = 5 > 0 ⇒ the graph is concave up at x = 3.

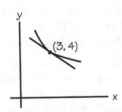

Find the line that passes through (3,4) with a slope of m = -1/2:

$$y - 4 = -1/2(x-3),$$
$$2y - 8 = -x + 3,$$
$$x + 2y = 11 \text{ is the tangent line.}$$

19.

	f	f'	f"
A	POS	POS	NEG
B	0	NEG	0
C	NEG	0	POS

25. f(x) has a relative maximum at values of x such that:

$$f'(x) = 0, \text{ and } f''(x) < 0.$$

The graph which is shown is that of f'(x) and we see that at x = 3 both the above conditions are satisfied: i.e., f'(3) = 0 and f"(3) < 0 (f'(x) has negative slope when x = 3, thus at this point there is a relative maximum).

31. f(6) = 3 means (6,3) on the graph. The slope at x = 6 is f'(6) = 2 (from Fig. 19). So the equation of the tangent line at (6,3) is:

$$y - 3 = 2(x - 6).$$

37. $f(x) = (3x^2 + 1)^4$,

$$f'(x) = 4(3x^2 + 1)^3 \cdot 6x = 24x(3x^2 + 1)^3.$$

Since $24x(3x^2 + 1)^3$ is always positive for x > 0, f'(x) is always positive. Therefore, the graph of f(x) is always increasing, and so II could not be the graph.

43. (a) Since the value of f'(9) is -1, a negative number, f(x) is decreasing at x = 9.

(b) f'(x) has a zero at x = 2, is positive to the left of x = 2, and negative to the right of x = 2, f(x) has a relative maximum at x = 2. The coordinates of the relative maximum point are (2,9).

(c) f'(x) has a zero at x = 10, is negative to the left of x = 10, and positive to the right of x = 2, f(x) has a relative minimum at x = 10. The coordinates of the relative minimum point are (10,1).

(d) f"(2) = -1, a negative number. Therefore, f(x) is concave down at x = 2.

(e) f"(x) is zero when x = 7 and it changes sign there. Hence, f(x) has an inflection point at x = 7. Since f(7) is 3, the coordinates of the inflection point are (7,3).

(f) f(x) is increasing at the rate of 6 units per unit change in x when f'(x) = 6. To solve f'(x) = 6, start at 6 on the y-axis, move right to the graph of y = f'(x), and move down to the x-axis to x = 15. Since f'(15) = 6, x = 15 is the answer.

2.3 Curve Sketching (Introduction)

The curves sketched in this section will look like one of the curves shown on the right or one of these curves turned upside down. The first type of curve is called a *parabola* and is the graph of a quadratic function. The second curve is called a *cubic* and is the graph of a certain type of cubic polynomial. The purpose of this section is to apply the first and second derivative rules to find the graphs of functions. Therefore, you should not automatically assume that the graph will be one of those on the right, but should reason its shape from the two derivative rules. For instance, every time that you claim that a point is an extreme point, you should determine the value of the second derivative at that point and draw your conclusion from the second derivative test. Later in the text, we will graph functions about which we have no prior information.

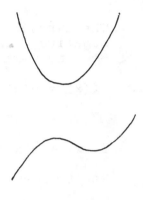

1. $f(x) = 2x^2 - 8$,

    $f'(x) = 4x$,

    $f''(x) = 4$.

    Set $f'(x) = 0$ and solve for x. If $4x = 0$ then $x = 0$. Substitute this value for x back into $f(x)$ to find the y value for this extreme point.

    $$f(0) = 2(0)^2 - 8 = -8.$$

    The extreme point is $(0,-8)$. Since $f''(0) = 4$, which is positive, the graph of $f(x)$ is concave up at $x = 0$. A graph of this function is shown in the answer section of the text.

7. $f(x) = -x^2 - 8x - 10$,

    $f'(x) = -2x - 8$,

    $f''(x) = -2$.

    Set $f'(x) = 0$ and solve for x:

    $$-2x - 8 = 0,$$
    $$-2x = 8,$$
    $$x = -4.$$

    Substitute this value for x back into $f(x)$ to find the y value for this extreme point.

$$f(-4) = -(-4)^2 - 8(-4) - 10$$
$$= -16 + 32 - 10 = 6.$$

The extreme point is $(-4,6)$. Since $f''(-4) = -2$, which is negative, the graph of $f(x)$ is concave down at $x = -4$. See the text's answer section for the graph of $f(x)$.

13. $f(x) = -\frac{1}{9}x^3 + x^2 + 9x$,

   $f'(x) = -\frac{1}{3}x^2 + 2x + 9$,

   $f''(x) = -\frac{2}{3}x + 2$.

Set $f'(x) = 0$ and solve for x:

$$-\frac{1}{3}x^2 + 2x + 9 = 0.$$

Multiplying both sides of the equation by -3, we obtain

$$x^2 - 6x - 27 = 0,$$
$$(x + 3)(x - 9) = 0,$$
$$x = -3, \text{ or } x = 9.$$

Substitute these values of x back into $f(x)$ to find the y values for these possible relative extreme points:

$$f(-3) = -\frac{1}{9}(-3)^3 + (-3)^2 + 9(-3) = 3 + 9 - 27 = -15.$$
$$f(9) = -\frac{1}{9}(9)^3 + (9)^2 + 9(9) = -81 + 81 + 81 = 81.$$

The possible extreme points are $(-3,-15)$ and $(9,81)$. Now

$$f''(-3) = -\frac{2}{3}(-3) + 2 = 2 + 2 = 4 > 0.$$
$$f''(9) = -\frac{2}{3}(9) + 2 = -6 + 2 = -4 < 0.$$

The graph of $f(x)$ is concave up at $x = -3$ and concave down at $x = 9$, hence $(-3,-15)$ is a relative minimum point and $(9,81)$ is a relative maximum point. See the graph in the answer section of the text.

19. $y = 1 + 3x^2 - x^3$,

   $y' = 6x - 3x^2$,

   $y'' = 6 - 6x$.

Set $y' = 0$ and solve for x:

$$6x - 3x^2 = 0, \quad x(6 - 3x) = 0,$$
$$x = 0, \text{ or } x = 2.$$

Use these values for x to find the y values of the possible relative extreme points:

At $x = 0$, $y = 1 + 3(0)^2 - (0)^3 = 1$.

At $x = 2$, $y = 1 + 3(2)^2 - (2)^3 = 1 + 12 - 8 = 5$.

The possible relative extreme points are $(0,1)$ and $(2,5)$. Next, check the second derivative.

At $x = 0$, $y'' = 6 - 6(0) = 6$ (graph concave up).

At $x = 2$, $y'' = 6 - 6(2) = -6$ (graph concave down).

So the function has a relative minimum at $x = 0$ and a relative maximum at $x = 2$. Since the concavity reverses somewhere between $x = 0$ and $x = 2$, there must be at least one inflection point. To find this point set $y'' = 0$:

$$6 - 6x = 0,$$

$$x = 1.$$

Substitute $x = 1$ into the function to find the y value for this inflection point.

$$y = 1 + 3(1)^2 - (1)^3 = 1 + 3 - 1 = 3.$$

The inflection point is $(1,3)$. See the graph in the answer section of the text.

25. $f(x) = ax^2 + bx + c$, $a \neq 0$,
$f'(x) = 2ax + b$,
$f''(x) = 2a$.

Since $a \neq 0$, $f''(x)$ is never zero. Since the second derivative of a function must be zero at an inflection point, $f(x)$ has no inflection points.

31. $f(x) = 5x^2 + x - 3$,
$f'(x) = 10x + 1$,
$f''(x) = 10$.

Set $f'(x) = 0$ and solve for x:

$$10x + 1 = 0,$$

$$10x = -1,$$

$$x = -.1.$$

Substitute this value for x back into $f(x)$ to find the y value for this relative extreme point.

$$f(-.1) = 5(-.1)^2 + (-.1) - 3 = .05 - .10 - 3.00 = -3.05$$

Since $f''(-.1) = 10$, which is positive, the graph is concave up at $x = -.1$ and hence $(-.1, -3.05)$ is a relative minimum point.

37. (a) Since 1990 is 10 years after 1980, the answer is f(10), which is 70. Therefore, 70% of homes had a VRC in 1990.

(b) We must solve f(t) = 10 for t. Start at 10 on the y-axis, move to the right to the graph of y = f(t), and move down to reach 4 on the t-axis. Therefore, f(4) = 10. So 10% of the homes had a VCR in 1984.

(c) The answer is f'(4), which is 5. Therefore, in 1984 the percentage of homes with VCRs was growing at the rate of 5% per year.

(d) We must solve f'(t) = 5, where t corresponds to a year in the second half of the 1980s. The horizontal line y = 5 crossed the graph of y = f'(x) twice, at t = 4 and at t = 10. The proper answer is 1990 (t = 10).

(e) The percentage of VCRs are growing at the greatest rate where f'(t) has a maximum, that is, where f"(t) is 0. This occurs in 1987 (t = 7). Since, f(7)= 40, 40% of the homes had VCRs in 1987.

2.4 Curve Sketching (Conclusion)

The two most important types of curves that are graphed in this section are the cubics without relative extreme points (such as the first curve or the upside down version of this curve) and the curves with asymptotes. These curves are stressed since, along with the curves considered in Section 2.3, they occur frequently in applications. However, there are a few other types of curves that occur in the exercises. As previously, all reasoning should be based on the first and second derivative tests.

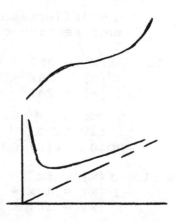

1. To find the x-intercepts of the function $y = x^2 - 3x + 1$, use the quadratic formula to find the values of x for which f(x) = 0:

$$y = x^2 - 3x + 1 \quad (a = 1, \ b = -3, \ c = 1),$$

$$b^2 - 4ac = (-3)^2 - 4 \cdot 1 \cdot 1 = 5,$$

$$x = \frac{-(-3) \pm \sqrt{5}}{2 \cdot 1} = \frac{3 \pm \sqrt{5}}{2}.$$

Therefore, the x-intercepts are $\left(\frac{3+\sqrt{5}}{2}, 0\right)$ and $\left(\frac{3-\sqrt{5}}{2}, 0\right)$.

7. If $f(x) = \frac{1}{3}x^3 - 2x^2 + 5x$, then

$$f'(x) = x^2 - 4x + 5.$$

If we apply the quadratic formula to $f'(x)$ with $a = 1$, $b = -4$, and $c = 5$, we find that $b^2 - 4ac = (-4)^2 - 4 \cdot 1 \cdot 5 = -4$, a negative number. Since we cannot take the square root of a negative number, there are no values of x for which $f'(x) = 0$. Hence $f(x)$ has no relative extreme points.

13. $f(x) = 5 - 13x + 6x^2 - x^3,$

$f'(x) = -13 + 12x - 3x^2,$

$f''(x) = 12 - 6x.$

If we apply the quadratic formula to $f'(x)$, we find that

$$b^2 - 4ac = (12)^2 - 4(-3)(-13) = 144 - 156 = -12,$$

which is negative. Since we cannot take the square root of a negative number, there are no values of x for which $f'(x) = 0$. So $f(x)$ has no relative extreme points. If we evaluate $f'(x)$ at some x, say $x = 0$, we see that the first derivative is negative, and so $f(x)$ is decreasing there. Since the graph of $f(x)$ is a smooth curve with no relative extreme points and no breaks, $f(x)$ must be decreasing for all x. Now, to check the concavity of $f(x)$, we must find where $f''(x)$ is negative, positive, or zero.

$f''(x) = 12 - 6x = 6(2 - x)$ is:

positive if $x < 2$,   (graph is concave up)
negative if $x > 2$,   (graph is concave down)
zero if $x = 2$.   (concavity reverses)

The inflection point is $(2, f(2)) = (2, -5)$. The y-intercept is $(0, f(0)) = (0, 5)$. To further improve the sketch of the graph, first sketch the tangent line at the inflection point. To do this compute the slope of the graph at $(2, -5)$.

$$f'(2) = -13 + 12(2) - 3(2)^2 = -1.$$

To sketch the graph, plot the inflection point and the y-intercept, draw the tangent line at $(2, -5)$, and then draw a curve that has this tangent line and is decreasing for all x, is concave down for $x > 2$, and concave up for $x < 2$. See the graph in the answer section of the text.

19. $f(x) = x^4 - 6x^2$,

$f'(x) = 4x^3 - 12x$,

$f''(x) = 12x^2 - 12$.

To find possible relative extreme points, set $f'(x) = 0$ and solve for x:

$$4x^3 - 12x = 0,$$
$$4x(x^2 - 3) = 0,$$
$$4x(x - \sqrt{3})(x + \sqrt{3}) = 0,$$
$$x = 0, \text{ or } x = \pm\sqrt{3}.$$

Substitute these values into $f(x)$ to find the y values for these possible relative extreme points:

$$f(0) = (0)^4 - 6(0)^2 = 0,$$
$$f(\sqrt{3}) = (\sqrt{3})^4 - 6(\sqrt{3})^2 = -9,$$
$$f(-\sqrt{3}) = (-\sqrt{3})^4 - 6(-\sqrt{3})^2 = -9.$$

The points are $(0,0)$, $(\sqrt{3},-9)$, and $(-\sqrt{3},-9)$.

Now $f''(0) = 12(0)^2 - 12 = -12$. Therefore, $f(x)$ is concave down at $x = 0$ and $(0,0)$ is a relative maximum point. For the other two points, we compute

$$f''(-\sqrt{3}) = 12(-\sqrt{3})^2 - 12 = 12(3) - 12 = 24,$$
$$f''(\sqrt{3}) = 12(\sqrt{3})^2 - 12 = 12(3) - 12 = 24.$$

Hence $f(x)$ is concave up at $x = -\sqrt{3}$ and $x = \sqrt{3}$, and so $(\sqrt{3},-9)$ and $(-\sqrt{3},-9)$ are relative minimum points. The concavity of this function reverses twice, so there must be at least two inflection points. To find these set $f''(x) = 0$ and solve for x:

$$12x^2 - 12 = 0,$$
$$x^2 - 1 = 0,$$
$$(x - 1)(x + 1) = 0, \text{ or } x = \pm 1.$$

The corresponding y-values are given by

$$f(1) = (1)^4 - 6(1)^2 = -5,$$
$$f(-1) = (-1)^4 - 6(-1)^2 = -5.$$

The inflection points are $(1,-5)$ and $(-1,-5)$. See the graph in the answer section of the text.

25. $y = \dfrac{9}{x} + x + 1,$

$y' = -\dfrac{9}{x^2} + 1,$

$y'' = \dfrac{18}{x^3}.$

Set $y' = 0$ and solve for x:

$$-\dfrac{9}{x^2} + 1 = 0,$$

$$1 = \dfrac{9}{x^2} \quad \text{(Multiply both sides by } x^2.)$$

$$x^2 = 9,$$

$$x = 3. \quad \text{(Note: We are only considering } x > 0.)$$

When $x = 3$, $y = \dfrac{9}{3} + 3 + 1 = 7,$ and

$$y'' = \dfrac{18}{3^3} > 0 \quad \text{(concave up)}.$$

Therefore, $(3,7)$ is a relative minimum point. Since y'' can never be 0, there are no inflection points. The term $\dfrac{9}{x}$ in the function y tells us that the y-axis is an asymptote. Also, as x gets large, the graph of y gets arbitrarily close to the straight line $y = x + 1$. Therefore, $y = x + 1$ is an asymptote of the graph. See the graph in the answer section of the text.

Helpful Hint: In general, $f'(a) = 0$ does not necessarily mean that $f(x)$ has an extreme point at $x = a$. Also, $f''(a) = 0$ does not necessarily mean that $f(x)$ has an inflection point at $x = a$. However, if $f(x)$ is quadratic or of the form $ax + \dfrac{b}{x} + c$, then $f'(a) = 0$ guarantees an extreme point at $x = a$. (Neither of these two types of functions has inflection points.) If $f(x)$ is a cubic polynomial, then there is one inflection point and it occurs where the second derivative is zero. (*Note:* These observations will help you in checking your work. On exams, you must still show that extreme and inflection points have the properties you claim.)

31. $g(x) = f'(x)$,
because $f(x)$ has 2 relative minima and a relative maximum and $g(x) = 0$ for these values of x.

$f(x) \neq g'(x)$.
To see this, look at the relative maximum of $g(x)$. Since $f(x)$ is not zero for this value of x, $f(x) \neq g'(x)$.

2.5 Optimization Problems

The procedure for solving an optimization problem can be thought of as consisting of the following two primary parts.

    (a) Find the function to be minimized.
    (b) Make a rough sketch of the graph of the function.

Part (a) is new to this section and requires careful reading of the problem. Part (b) relies on techniques presented in Sections 2.3 and 2.4. However, certain shortcuts can be taken when sketching curves in this section. For instance, if we only want to find the x-coordinate of a relative extreme point, we can make a very rough estimate of the y-coordinate when sketching the curve.

    One of the common mistakes that students make when working optimization problems on exams is to just set the first derivative equal to 0 and solve for x. They forget to apply the second derivative test at the value of x found.

1. $g(x) = 10 + 40x - x^2$,
      $g'(x) = 40 - 2x$,
      $g''(x) = -2$.

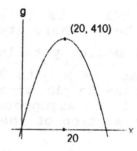

     The graph is obtained by the curve-sketching technique of Section 2.3.

     Therefore, the maximum value of $g(x)$ occurs at $x = 20$.

7.   **a.** See drawing in answer section of the text.

     b. The girth of the box is given by $x + x + x + x = 4x$. The equation for the length plus the girth is

$$h + 4x.$$

     c. The equation for volume, which we wish to maximize, is the objective equation,

$$V = x^2 h.$$

     The constraint equation is

$$h + 4x = 84,$$
     or
$$h = 84 - 4x.$$

     d. Substituting $h = 84 - 4x$ into the objective equation, we have

$$V = x^2(84 - 4x) = -4x^3 + 84x^2.$$

e. By making a rough sketch of the graph of $V = -4x^3 + 84x^2$, we can find the value of x corresponding to the greatest volume.

$$V' = -12x^2 + 168x,$$

$$V'' = -24x + 168.$$

Solve $V' = 0$:

$$-12x^2 + 168x = 0,$$

$$x^2 - 14x = 0,$$

$$x(x - 14) = 0,$$

$$x = 0, \text{ or } x = 14.$$

V'' is positive at $x = 0$ and negative at $x = 14$. Therefore V has a relative minimum at $x = 0$ and a relative maximum at $x = 14$. When $x = 0$, $V = 0$. When $x = 14$, $V = -4(14)^3 + 84(14)^2 = 5488$. The sketch of the graph shows that the maximum value of V occurs when $x = 14$. From the constraint equation, $h = 84 - 4(14) = 28$.

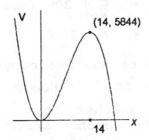

(14, 5844)

14

13. The problem asks that total area be maximized. Let A be the area, let x be the length of the fence parallel to the river, and let y the length of each section perpendicular to the river. The objective equation gives an expression for A in terms of the other variables:

$$A = xy \qquad \text{(Objective equation)}.$$

The cost of the fence parallel to the river is 6x dollars (x feet at $6 per foot) and the cost of the three sections perpendicular to the river is 3·5y or 15y dollars. Since $1500 is available to build the fence, we must have

$$6x + 15y = 1500 \qquad \text{(Constraint equation)}.$$

Solving for y,

$$15y = -6x + 1500,$$

$$y = -\frac{2}{5}x + 100.$$

2-13

Substituting this expression for y into the objective equation, we obtain

$$A = x\left(-\frac{2}{5}x + 100\right) = -\frac{2}{5}x^2 + 100x.$$

The graph of $A = -\frac{2}{5}x^2 + 100x$ is easily obtained by our curve sketching techniques. The maximum value occurs when $x = 125$. Substituting into $y = -\frac{2}{5}x + 100$, this gives us $y = -\frac{2}{5}(125) + 100 = 50$. Therefore, the optimum dimensions are $x = 125$, $y = 50$.

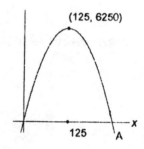

19. $x = 20 - \frac{w}{2}$ (from the constraint equation).

$A = wx$ (the objective function).

Therefore:

$$A = w\left(20 - \frac{w}{2}\right) = -\frac{w^2}{2} + 20w \text{ (has a parabola shape)},$$

$$\frac{dA}{dw} = -\frac{2w}{2} + 20 = -w + 20.$$

A relative minimum or maximum occurs when $\frac{dA}{dw} = 0$. Hence $-w + 20 = 0 \Rightarrow w = 20$. (Since $x = 20 - \frac{w}{2}$, $x = 10$.) The next question is whether this is a relative maximum. Thus, looking at the second derivative:

$$\frac{d^2A}{dw^2} = -1 < 0,$$

this means that $w = 20$ is a relative max. To sketch the graph of A we see that $A = 0$ when $w = 0$ and when $x = 0$: $20 - \frac{w}{2} = 0 \Rightarrow \frac{w}{2} = 20$, or $w = 40$. At $w = 20$,

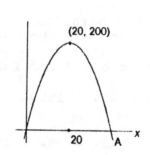

$$A = 20\left(20 - \frac{20}{2}\right)$$
$$= 200.$$

2.6 Further Optimization Problems

The exercises in this section are of four types.

(a) Variations of the types of exercises considered in Section 2.5.

(b) Inventory problems. These problems have the same constraint equation, r·x = [number of items used or manufactured during the year], and similar objective equations, [costs] = ...

(c) Orchard problems (that is, problems similar to the first practice problem or to Example 1). These problems have the same type of objective equation, [profit or revenue] = [amount of money]·[quantity]. The constraint equations are found by reading the problem to see if [amount of money] depends on [quantity] or vice versa, and expressing the dependent variable in terms of the other.

(d) Original problems. Situations that use the same type of machinery as the examples worked in the text but require a fresh approach.

1. The quantity to be maximized is revenue, call it R. Since the problem asks for the optimum number of prints, let x be the number of prints to be offered for sale. The other quantity that varies is price per print, p. The objective equation is

$$[revenue] = [price per print]·[number of prints],$$

$$R = p·x.$$

Since the price per print depends on the number of prints offered for sale, the constraint equation can be obtained by expressing p in terms of x.

$$[price per print] = \begin{bmatrix} original\ price \\ per\ print \end{bmatrix} - \begin{bmatrix} decrease\ in\ price \\ (per\ print)\ due \\ to\ increase \end{bmatrix}.$$

Thus

$$[price per print] = 400 - (x - 50)·5$$
$$= 400 - 5x + 250$$
$$= 650 - 5x.$$

The decrease in price (per print) due to the increase in the number of prints was obtained by multiplying x - 50, the number of prints in excess of 50, by the price reduction for each excess print. Therefore,

$$R = px = (650 - 5x)x = 650x - 5x^2.$$

The graph of R is found by curve sketching. Its maximum value occurs when $x = 65$. Therefore, 65 prints should be offered for sale.

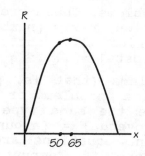

7. We let r represent the number of production runs, and x the number of microscopes manufactured per run. Hence the total number of microscopes manufactured is

$$\left[\begin{array}{c} \text{number of microscopes} \\ \text{per production run} \end{array}\right] \cdot \left[\begin{array}{c} \text{number} \\ \text{of runs} \end{array}\right] = xr.$$

Hence we have as a constraint equation

$$xr = 1600, \text{ or } x = \frac{1600}{r}.$$

We wish to minimize inventory expenses. There are three expenses which make up the total cost. The storage costs (SC), based on the maximum number of microscopes in the warehouse, will be

$$SC = 15x,$$

since the warehouse is the most full just after a run, which will produce x microscopes. Since the insurance costs are based on the average number of microscopes in the warehouse, and the average number of microscopes is given by $\frac{x}{2}$, the insurance costs (IC) are

$$IC = 20\left(\frac{x}{2}\right).$$

Each production run costs $2500, so the total production costs (PC) are given by

$$PC = 2500r.$$

Hence the objective equation is

$$C = SC + IC + PC$$

$$= 2500r + 15x + 20\left(\frac{x}{2}\right)$$

$$= 2500r + 25x.$$

Substituting $x = \frac{1600}{r}$ into the objective equation we have,

$$C = 2500r + 25\left(\frac{1600}{r}\right)$$

$$= 2500r + \frac{40,000}{r}.$$

We use our curve sketching techniques to make a rough sketch of this function. Only positive values of r are relevant; that is, $r > 0$.

$$C' = 2500 - \frac{40,000}{r^2},$$

$$C'' = \frac{80,000}{r^3}.$$

Set $C' = 0$ and solve for r:

$$2,500 - \frac{40,000}{r^2} = 0,$$

$$2,500 = \frac{40,000}{r^2},$$

$$r^2 = \frac{40,000}{2,500} = 16, \text{ or } r = 4.$$

When $r = 4$, $C'' > 0$. There is a relative minimum at $r = 4$. Therefore, there should be 4 production runs.

Note: The value of C when $r = 4$ need not be calculated. After establishing the existence of a relative minimum point when $r = 4$, we can use our familiarity with graphs of functions of the form $y = \frac{a}{x} + bx$ to make a rough sketch.

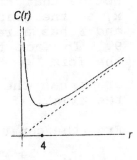

13. Let w and x be the dimensions of the corral as shown in the text. The corral is to have an area of 54 square meters, hence the constraint equation is

$$xw = 54 \text{ or } w = \frac{54}{x} \qquad \text{(Constraint)}.$$

The total amount of fencing needed is two pieces of length x and three pieces of length w. Hence the objective equation is

$$F = 2x + 3w \qquad \text{(Objective)}.$$

Substituting $w = \frac{54}{x}$ into the objective equation we have,

$$F = 2x + 3\left(\frac{54}{x}\right) = 2x + \frac{162}{x}.$$

Using curve sketching techniques, we make a rough sketch of the graph.

$$F = 2x + \frac{162}{x}, \quad x > 0,$$

$$F' = 2 - \frac{162}{x^2},$$

$$F'' = \frac{324}{x^3}.$$

Set $F' = 0$ and solve for x:

$$2 - \frac{162}{x^2} = 0,$$

$$2 = \frac{162}{x^2},$$

$$x^2 = \frac{162}{2} = 81, \text{ or } x = 9.$$

Notice that $F'' > 0$ for all positive x, so the graph of F is concave up, and F has a relative minimum at x = 9. In fact the formula for F has the form $y = \frac{a}{x} + bx$, and so the graph has the basic shape shown at the right.

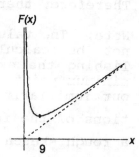

From the constraint equation we find

$$w = \frac{54}{x} = \frac{54}{9} = 6.$$

Therefore, w = 6, x = 9 are the optimum dimensions.

19. First consider the diagram at the right. Let x be the length of each edge of the square ends and h the other dimension. Since the sum of the three dimensions can be at most 120 centimeters, the constraint equation is

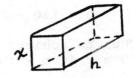

$$2x + h = 120, \text{ or } h = 120 - 2x.$$

Since we wish to maximize the volume of the package, the objective equation is

$$V = x^2 h.$$

Substituting h = 120 − 2x into the objective equation, we have

$$V = x^2(120 - 2x) = 120x^2 - 2x^3,$$

$$V' = 240x - 6x^2,$$

$$V'' = 240 - 12x.$$

Set V′ = 0 and solve for x:

$$240x - 6x^2 = 0,$$

$$6x(40 - x) = 0,$$

$$x = 0, \text{ and } x = 40.$$

Check concavity: At x = 0,

$$V'' = 420 - 12(0) = 240 > 0 \qquad \text{(graph concave up)}.$$

At x = 40,

$$V'' = 240 - 12(40) = -240 < 0 \quad \text{(graph concave down)}.$$

Therefore, the graph has a relative minimum at x = 0 and a relative maximum at x = 40. When x = 0, V = 0; at x = 40,

$$V = 120(40)^2 - 2(40)^3 = 64,000.$$

A rough sketch of the graph

$$V = 120x^2 - 2x^3, \quad x \geq 0,$$

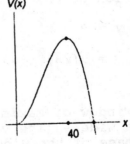

reveals a maximum value at x = 40.   The value of h corresponding to x = 40 is found from the constraint equation

$$h = 120 - 2x = 120 - 2(40) = 40.$$

Thus to achieve maximum volume, the package should be 40cm × 40cm × 40cm.

25. First, consider the diagram at the right. The base of the window has length 2x. The area of the window is given by

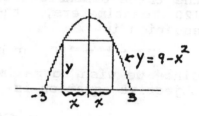

$$A = 2xy = 2x(9 - x^2)$$
$$= 18x - 2x^3.$$

We now make a rough sketch of the graph of $A = 18x - 2x^3$, for x > 0:

$$A' = 18 - 6x^2,$$
$$A'' = -12x.$$

Set A' = 0:

$$18 - 6x^2 = 0,$$
$$18 = 6x^2,$$
$$3 = x^2, \text{ or } x = \sqrt{3}.$$

When $x = \sqrt{3}$, $A'' = -12(\sqrt{3})$ is negative. Therefore the graph of $A = 18x - 2x^3$ has a relative maximum value at $x = \sqrt{3}$. When $x = \sqrt{3}$,

$$A = 18\sqrt{3} - 2(\sqrt{3})^3$$
$$= 18\sqrt{3} - 2 \cdot 3 \cdot \sqrt{3} \text{ [since } (\sqrt{3})^3$$
$$= (\sqrt{3} \cdot \sqrt{3}) \cdot \sqrt{3} = 3\sqrt{3}]$$
$$= 12\sqrt{3}.$$

A rough sketch of the graph reveals a maximum value at $x = \sqrt{3}$. The height y, corresponding to $x = \sqrt{3}$ is found from the equation of the parabola

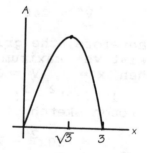

$$y = 9 = x^2$$
$$= 9(\sqrt{3})^2 = 9 - 3 = 6.$$

The window of maximum area should be 6 units high and $2\sqrt{3}$ units wide. (The value of x gives only half the width of the base of the window.)

2.7 Applications of Calculus to Business and Economics

This section, which explores cost, revenue, and profit functions, and applies calculus to the optimization of profits of a firm, is independent of the rest of the text.

1. If the cost function is

$$C(x) = x^3 - 6x^2 + 13x + 15,$$

the marginal cost function is

$$M(x) = C'(x) = 3x^2 - 12x + 13.$$

We find the minimum value of $M(x)$ by making a rough graph, a simple matter since $M(x)$ is a quadratic function.

$$M'(x) = 6x - 12,$$
$$M''(x) = 6.$$

Set $M'(x) = 0$ to locate the x where the marginal cost is minimized.

$$6x - 12 = 0$$
$$6x = 12, \text{ or } x = 2.$$

Now, $M(2) = 3(2)^2 - 12(2) + 13 = 1$ and $M''(2)$ is positive. Therefore, $M(x)$ has a relative minimum at $x = 2$. A sketch of the graph reveals a minimum value of 1. That is, the minimum marginal cost is \$1.

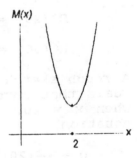

Note: Sketching the graph is not absolutely necessary. Once we realize that it is a parabola opening upward, we know that its minimum value will occur where the first derivative is zero.

7. If the demand equation for the commodity is given by $p = \frac{1}{12}x^2 - 10x + 300$, $0 \le x \le 60$, then the revenue function is

$$R(x) = x \cdot p = x\left(\frac{1}{12}x^2 - 10x + 300\right) = \frac{1}{12}x^3 - 10x^2 + 300x.$$

To find the maximum value of $R(x)$ we first make a rough sketch of its graph.

$$R(x) = \frac{1}{12}x^3 - 10x^2 + 300x,$$

$$R'(x) = \frac{1}{4}x^2 - 20x + 300,$$

$$R''(x) = \frac{1}{2}x - 20.$$

Set $R'(x) = 0$ and solve for x:

$$\frac{1}{4}x^2 - 20x + 300 = 0,$$

$$x^2 - 80x + 1200 = 0, \quad \text{(multiply by 4)}$$

$$(x - 20)(x - 60) = 0,$$

$$x = 20 \text{ or } x = 60.$$

[Note: To see how to factor $x^2 - 80x + 1200$, first try to factor $x^2 - 8x + 12$.]
Now,

$$R(20) = \frac{1}{12}(20)^3 - 10(20)^2 + 300(20) = 2666 \ 2/3,$$

$$R(60) = \frac{1}{12}(60)^3 - 10(60)^2 + 300(60) = 0,$$

$$R''(20) = \frac{1}{2}(20) - 20 = -10 < 0,$$

$$R''(60) = \frac{1}{2}(60) - 20 = 10 > 0.$$

A rough sketch of the graph reveals that revenue is maximized
when $x = 20$. From the demand
equation,

$$p = \frac{1}{12}(20)^2 - 10(20) + 300$$

$$= 133 \ 1/3.$$

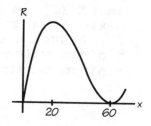

13. **a.** Since [profit] = [revenue] − [cost], and since revenue
   is given by the function

$$R(x) = x \cdot p = 60x - (10^{-5})x^2,$$

the profit function is

$$P(x) = R(x) - C(x) = (60x - (10^{-5})x^2) - (7 \cdot 10^6 + 30x)$$

$$= 30x - (10^{-5})x^2 - 7 \cdot 10^6$$

$$= -(10^{-5})x^2 + 30x - 7 \cdot 10^6.$$

The graph of $P(x)$ is a parabola opening downward [since

the coefficient of x^2 is $-(10^{-5})$, which is negative] and has its maximum value at its unique relative maximum point. Therefore, it is sufficient to solve $P'(x) = 0$ for x.

$$P'(x) = -2(10^{-5})x + 30.$$

Set $P'(x) = 0$:

$$-2(10^{-5})x + 30 = 0,$$

$$-2(10^{-5})x = -30,$$

$$x = \frac{-30}{-2(10^{-5})} = 15 \cdot 10^5.$$

At $x = 15 \cdot 10^5$ the corresponding price for a thousand kilowatt hours is

$$p = 60 - (10^{-5})(15 \cdot 10^5) = \$45.$$

b. The new profit function is given by

$$P_1(x) = R(x) - C_1(x) = 60x - (10^{-5})x^2 - (7 \cdot 10^6 + 40x)$$

$$= 20x - (10^{-5})x^2 - 7 \cdot 10^6.$$

To find the maximum value for this function we find $P_1'(x)$ and then set $P'(x) = 0$ and solve for x:

$$P_1'(x) = 20 - 2 \cdot 10^{-5}x,$$

$$20 - 2 \cdot 10^{-5}x = 0,$$

$$2 \cdot 10^{-5}x = 20, \text{ or } x = 10^6.$$

At $x = 10^6$ the corresponding price for a thousand kilowatt hours is

$$p = 60 - (10^{-5})(10^6) = \$50.$$

So the answer is no, the company should not pass all of the \$10 increase on to the consumer if it wants to maximize profit. With the new cost function the price that maximizes profit is \$50, reflecting a \$5 increase.

Chapter 2: Supplementary Exercises

The most difficult problems in this chapter are the optimization problems. These problems cannot be solved all at once, but require that you break them down into small manageable pieces. Although some of them fit into neat categories, others require patient analysis.

1. Graph b, because the population is growing at an increasing rate.

7. $f'(x)$ is positive at a and c in the sense that it is non-negative $f'(x) = 0$ at these points.
    $f'(x)$ is positive at d and e and is non-zero at these points.

13. Since $f(1) = 2$, the graph goes through the point $(1,2)$.
    Since $f'(1) > 0$, the graph is increasing at $x = 1$.

19. Since $g(5) = -1$, the graph goes through the point $(5,-1)$.
    Since $g'(5) = -2 < 0$, the graph is decreasing at $x = 5$.
    The fact that $g''(5) = 0$ is inconclusive since it is possible for $g''(x) = 0$ without x being an inflection point.
    For example, if $g(x)$ is a straight line, $g''(x) = 0$ for all x but $g(x)$ has no inflection points.

25. If $y = x^2 + 3x - 10$, then
    $$y' = 2x + 3,$$
    $$y'' = 2.$$

    Set $y' = 0$ and solve for x to find the relative extreme point:
    $$2x + 3 = 0,$$
    $$x = -\frac{3}{2}.$$

    If $x = -\frac{3}{2}$, then $y = \left(-\frac{3}{2}\right)^2 + 3\left(-\frac{3}{2}\right) - 10 = \frac{9}{4} - \frac{9}{2} - 10 = -\frac{49}{4}$.
    Since $y'' = 2 > 0$, the parabola is concave up and the point $\left(-\frac{3}{2}, -\frac{49}{4}\right)$ is a minimum. When $x = 0$, $y = 0^2 + 3(0) - 10 = -10$. Hence the y-intercept is $(0,-10)$. To find the x-intercepts, set $y = 0$ and solve for x:
    $$x^2 + 3x - 10 = 0,$$
    $$(x + 5)(x - 2) = 0,$$
    $$x = -5 \text{ and } x = 2.$$

    So the x-intercepts are $(-5,0)$ and $(2,0)$. See the graph in the answer section of the text.

31. If $y = -x^2 + 20x - 90$, then

$$y' = -2x + 20,$$

$$y'' = -2.$$

Set $y' = 0$ and solve for x to find the relative extreme point:

$$-2x + 20 = 0,$$

$$x = 10.$$

If $x = 10$, $y = -(10)^2 + 20(10) - 90 = -100 + 200 - 90 = 10$. Since $y'' = -2 < 0$, the parabola is concave down and $(10,10)$ is the relative maximum point. When $x = 0$, $y = -0^2 + 20(0) - 90 = -90$. Hence the y-intercept is $(0,-90)$. To find the x-intercept we use the quadratic formula:

$$b^2 - 4ac = (20)^2 - 4(-1)(-90) = 40,$$

$$\sqrt{b^2 - 4ac} = \sqrt{40} = \sqrt{4 \cdot 10} = \sqrt{4} \cdot \sqrt{10} = 2\sqrt{10}.$$

$$x = \frac{-20 \pm 2\sqrt{10}}{-2} = 10 \pm \sqrt{10}.$$

So the x-intercepts are $(10+\sqrt{10},0)$ and $(10-\sqrt{10},0)$. See the graph in the answer section of the text.

37. $y = \frac{11}{3} + 3x - x^2 - \frac{1}{3}x^3,$

$y' = 3 - 2x - x^2,$

$y'' = -2 - 2x.$

Set $y' = 0$ and solve for x to find the possible relative extreme points:

$$3 - 2x - x^2 = 0.$$

Multiply by -1 and rearrange terms.

$$x^2 + 2x - 3 = 0,$$

$$(x - 1)(x + 3) = 0,$$

$$x = 1, \text{ or } x = -3.$$

If $x = 1$, $y = \frac{11}{3} + 3(1) - (1)^2 - \frac{1}{3}(1)^3 = \frac{16}{3}$, and

$$y'' = -2 - 2 = -4 < 0.$$

Hence the graph is concave down and has a relative maximum at $\left(1, \frac{16}{3}\right)$. At $x = -3$, $y = \frac{11}{3} + 3(-3) - (-3)^2 - \frac{1}{3}(-3)^3 = -\frac{16}{3}$ and $y'' = -2 - 2(-3) = 4 > 0$. Hence the graph is concave up and has a relative minimum at $\left(-3, -\frac{16}{3}\right)$. To find the inflection point we set $y'' = 0$ and solve for x:

$$-2 - 2x = 0,$$
$$-2x = 2,$$
$$x = -1.$$

If $x = -1$, $y = \frac{11}{3} + 3(-1) - (-1)^2 - \frac{1}{3}(-1)^3 = 0$. Hence the inflection point is $(-1,0)$. See the graph in the answer section of the text.

43. This function has the form $y = g(x) + mx + b$, where $g(x) = \frac{20}{x}$ and $mx + b = \frac{x}{5} + 3$ $(= \frac{1}{5}x + 3$, a straight line of slope $1/5$).

$$y = \frac{x}{5} + \frac{20}{x} + 3, \qquad y' = \frac{1}{5} - \frac{20}{x^2}, \qquad y'' = \frac{40}{x^3}.$$

Set $y' = 0$ and solve for x:

$$\frac{1}{5} - \frac{20}{x^2} = 0, \qquad \frac{1}{5} = \frac{20}{x^2}, \qquad x^2 = 100.$$

(The last equation was obtained by cross multiplication.) Thus $x = 10$ (since only $x > 0$ is being considered). When $x = 10$, $y = \frac{10}{5} + \frac{20}{10} + 3 = 2 + 2 + 3 = 7$, and y'' is obviously positive. Therefore the graph is concave up at $x = 10$, and so $(10,7)$ is a relative minimum point. Since y'' can never be 0, there are no inflection points. The graph has the y-axis as a vertical asymptote and approaches the straight line $y = \frac{x}{5} + 3$ as x gets large. The graph is shown in the answer section of the text.

49. A and c, B and e, C and f, D and b, E and a, F and d.

55. First consider the figure. Let x be the width and h the height of the box. The box is to have a volume of 200 cubic feet, hence the constraint equation is

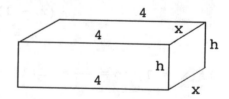

$$4xh = 200 \text{ or } h = \frac{50}{x}.$$

We minimize the amount of material by minimizing the amount of surface area. Our objective equation is

$$A = 4x + 2xh + 8h.$$

Substituting the value $h = \dfrac{50}{x}$ into the objective equation we have,

$$A = 4x + 2x\left(\frac{50}{x}\right) + 8\left(\frac{50}{x}\right)$$

$$= 4x + 100 + \frac{400}{x}.$$

We now make a rough sketch of the graph of this function.

$$A = 4x + 100 + \frac{400}{x}, \quad x > 0,$$

$$A' = 4 - \frac{400}{x^2},$$

$$A'' = \frac{800}{x^3}.$$

Set $A' = 0$ and solve for x:

$$4 - \frac{400}{x^2} = 0,$$

$$4 = \frac{400}{x^2},$$

$$x^2 = \frac{400}{4} = 100, \text{ or } x = 10.$$

When $x = 10$, $A = 4(10) + 100 + \dfrac{400}{10} = 180$ and $A'' = \dfrac{800}{10^3} > 0$.

The sketch of the function reveals a maximum area at $x = 10$. When $x = 10$, $h = \dfrac{50}{10} = 5$.

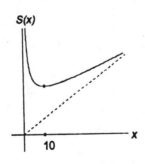

61. Since the profit function is

$$P(x) = R(x) - C(x),$$

and since

$$R(x) = x \cdot p = x(150 - .02x) = 150x - .02x^2,$$

we have

$$P(x) = (150x - .02x^2) - (10x + 300)$$
$$= 140x - .02x^2 - 300$$
$$= -.02x^2 + 140x - 300.$$

This graph of $P(x)$ is a parabola opening downward (since the coefficient of x^2 is negative) and therefore assumes its maximum value where the first derivative is zero.

$$P'(x) = -.04x + 140.$$

Setting $P'(x) = 0$ and solving for x we have

$$-.04x + 140 = 0$$

$$-.04x = -140$$

$$x = \frac{-140}{-.04} = 3500.$$

[Note: One way to evaluate $\frac{140}{.04}$ is to first multiply the numerator and denominator by 100 and then divide. That is, $\frac{140}{.04} = \frac{14,000}{4} = 3500.$]

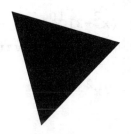

CHAPTER 3

TECHNIQUES OF DIFFERENTIATION

3.1 The Product and Quotient Rules

The first thirty exercises are routine drill. You will be able to tell from the answers if you are mastering the use of the product and quotient rules. In each case, try to obtain the simplified version of the answer shown in the answer section. Most students find this difficult, but now is a good time to practice a skill that may be important for the next exam!

1. To differentiate the product

$$(x+1)(x^3+5x+2)$$

let $f(x) = (x+1)$ and $g(x) = (x^3+5x+2)$ and apply the product rule. Then

$$\frac{d}{dx}[(x+1)(x^3+5x+2)] = (x+1) \cdot \frac{d}{dx}(x^3+5x+2) + (x^3+5x+2) \cdot \frac{d}{dx}(x+1)$$

$$= (x+1)(3x^2+5) + (x^3+5x+2)(1).$$

Carrying out the multiplication in the first product of the derivative and combining like terms we have

$$\frac{d}{dx}[(x+1)(x^3+5x+2)] = (3x^3+3x^2+5x+5) + (x^3+5x+2)$$

$$= 4x^3 + 3x^2 + 10x + 7.$$

7. To differentiate $(x^2+3)(x^2-3)^{10}$, let $f(x) = (x^2+3)$ and $g(x) = (x^2-3)^{10}$ and apply the product rule. When we compute $g'(x)$, we will need the general power rule.

$$\frac{d}{dx}[(x^2+3)(x^2-3)^{10}] = (x^2+3)\frac{d}{dx}(x^2-3)^{10} + (x^2-3)^{10}\cdot\frac{d}{dx}(x^2+3)$$

$$= (x^2+3)10(x^2-3)^9 2x + (x^2-3)^{10}2x.$$

If we factor $2x(x^2-3)^9$ out of each term of the derivative we have

$$\frac{d}{dx}[(x^2+3)(x^2-3)^{10}] = 2x(x^2-3)^9[(x^2+3)10 + (x^2-3)]$$

$$= 2x(x^2-3)^9(11x^2+27).$$

Helpful Hint: Here is a good way to remember the order of the terms in the quotient rule:

1. Draw a long fraction bar and write $[g(x)]^2$ in the denominator:

$$\frac{}{[g(x)]^2}$$

2. While you are still thinking about $g(x)$, write it in the numerator:

$$\frac{g(x)}{[g(x)]^2}$$

3. The rest is easy because each term in the numerator of the derivative involves one function and one derivative. Since $g(x)$ is already written, it must go with $f'(x)$. Next comes the minus sign and then the function and derivative that have not yet been used.

$$\frac{\mathbf{g(x)f'(x) - f(x)g'(x)}}{[g(x)]^2}$$

Think of $g(x)$ and $g'(x)$ as being "around the outside" of this formula.

13. To differentiate $\dfrac{x^2-1}{x^2+1}$, let $f(x) = x^2-1$ and $g(x) = x^2+1$ and apply the quotient rule:

$$\frac{d}{dx}\left(\frac{x^2-1}{x^2+1}\right) = \frac{(x^2+1)\cdot\frac{d}{dx}(x^2-1) - (x^2-1)\cdot\frac{d}{dx}(x^2+1)}{(x^2+1)^2}$$

$$= \frac{(x^2+1)2x - (x^2-1)2x}{(x^2+1)^2}.$$

If we do the multiplication in the numerator, this derivative simplifies.

$$\frac{d}{dx}\left(\frac{x^2-1}{x^2+1}\right) = \frac{2x^3 + 2x - 2x^3 + 2x}{(x^2+1)^2} = \frac{4x}{(x^2+1)^2}.$$

Helpful Hint: A problem such as #15 can be worked with the quotient rule, but since the numerator is a constant, the differentiation is easier if you rewrite the quotient with the notation $(\cdots)^{-1}$. See Practice Problem #2.

19. To differentiate $\dfrac{3x^2 + 5x + 1}{3 - x^2}$, let $f(x) = 3x^2 + 5x + 1$ and $g(x) = 3 - x^2$ and apply the quotient rule.

$$\frac{d}{dx}\left(\frac{3x^2 + 5x + 1}{3 - x^2}\right) = \frac{(3-x^2)\frac{d}{dx}(3x^2+5x+1) - (3x^2+5x+1)\frac{d}{dx}(3-x^2)}{(3-x^2)^2}$$

$$= \frac{(3-x^2)(6x+5) - (3x^2+5x+1)(-2x)}{(3-x^2)^2}.$$

If we do the multiplication in the numerator, we have

$$\frac{d}{dx}\left(\frac{3x^2 + 5x + 1}{3 - x^2}\right) = \frac{(18x+15-6x^3-5x^2) + (6x^3+10x^2+2x)}{(3-x^2)^2}$$

$$= \frac{5x^2 + 20x + 15}{(3-x^2)^2}$$

$$= \frac{5(x+1)(x+3)}{(3-x^2)^2}.$$

Warning: Some students prefer to work #19 using the product rule, but this may cause them more work than they realize. It is possible to write

$$y = \frac{3x^2+5x+1}{3-x^2} = (3x^2+5x+1)(3-x^2)^{-1},$$

and

$$\frac{dy}{dx} = (3x^2+5x+1)\cdot\frac{d}{dx}(3-x^2)^{-1} + (3-x^2)^{-1}\cdot\frac{d}{dx}(3x^2+5x+1)$$

$$= (3x^2+5x+1) \cdot (-1)(3-x^2)^{-2} \cdot (-2x) + (3-x^2)^{-1} \cdot (6x+5).$$

This method is not really shorter, but it does avoid the quotient rule. The main difficulty arises when you try to simplify the derivative, for you must factor out terms that involve negative exponents. However, if you really prefer this method, the next paragraph explains how to handle negative exponents.

How to simplify a sum involving negative exponents: Suppose that various powers of some quantity Q occur in each term of the sum. Find the exponent on Q that lies farthest to the left on the number line and factor out this power of Q from each term in the sum.

In the solution above, notice that $\frac{dy}{dx}$ involves $(3-x^2)^{-2}$ and $(3-x^2)^{-1}$. Since -2 is to the left of -1 on the number line, factor out $(3-x^2)^{-2}$. In the second term below we need $(3-x^2)^1$, because $(3-x^2)^{-1} = (3-x^2)^{-2}(3-x^2)^{+1}$.

$$\frac{dy}{dx} = (3-x^2)^{-2}[(3x^2+5x+1)(-1)(-2x) + (3-x^2)^1(6x+5)]$$

$$= (3-x^2)^{-2}[(6x^3+10x^2+2x) + (18x+15-6x^3-5x^2)]$$

$$= (3-x^2)^{-2}[5x^2+20x+15]$$

$$= 5(3-x^2)^{-2}(x+1)(x+3).$$

Moral of this story: If you have to simplify the derivative of a quotient, use the quotient rule for the differentiation step.

25. Simplify first:

$$\frac{x^4-4x^2+3}{x} = x^3-4x+\frac{3}{x}$$

$$= x^3-4x+3x^{-1}.$$

Then,

$$\frac{d}{dx}(x^3-4x+3x^{-1}) = 3x^2-4-3x^{-2}$$

$$= 3x^2-4-\frac{3}{x^2}$$

$$= \frac{3x^4-4x^2-3}{x^2}.$$

31. The solution to this exercise is a two-step procedure.

(a) Find the slope at the point where $x = 3$.

(b) Use the point-slope formula from Section 1.1 to obtain the equation of the line through the point $(3,16)$ having the slope found in (a).

Given $y = (x-2)^5(x+1)^2$, we have

$$\frac{dy}{dx} = (x-2)^5 \cdot \frac{d}{dx}(x+1)^2 + (x+1)^2 \cdot \frac{d}{dx}(x-2)^5 \quad \text{[Product Rule]}$$

$$= (x-2)^5 2(x+1) + (x+1)^2 5(x-2)^4 \quad \text{[Gen. Power Rule]}$$

Now, the slope of the curve at $x = 3$ is

$$(3-2)^5 2(3+1) + (3+1)^2 5(3-2)^4 = 8 + 5(16) = 88.$$

Hence, the equation of the tangent line at $(3,16)$ is

$$y - 16 = 88(x - 3).$$

37. To find the points on the graph of $y = \frac{x^2+3x-1}{x}$ where the slope is 5 we first take the derivative:

$$\frac{dy}{dx} = \frac{dy}{dx}\left(\frac{x^2+3x-1}{x}\right)$$

$$= \frac{x \cdot \frac{dy}{dx}(x^2+3x-1) - (x^2+3x-1) \cdot \frac{dy}{dx}(x)}{x^2} \quad \text{[Quotient Rule]}$$

$$= \frac{x(2x+3) - (x^2+3x-1)}{x^2}$$

$$= \frac{x^2 + 1}{x^2}.$$

Now, set the derivative equal to 5 and solve for x:

$$\frac{x^2 + 1}{x^2} = 5,$$

$$x^2 + 1 = 5x^2,$$

$$4x^2 = 1,$$

$$x^2 = \frac{1}{4}, \text{ or } x = \pm\frac{1}{2}.$$

Substituting $x = \frac{1}{2}$ and $x = -\frac{1}{2}$ into $y = \frac{x^2+3x-1}{x}$ we have

$$y = \frac{\left(\frac{1}{2}\right)^2 + 3\left(\frac{1}{2}\right) - 1}{\frac{1}{2}} = \frac{3}{2},$$

and,

$$y = \frac{\left(-\frac{1}{2}\right)^2 + 3\left(-\frac{1}{2}\right) - 1}{-\frac{1}{2}} = \frac{9}{2}.$$

The points where the slope is 5 are $\left(\frac{1}{2}, \frac{3}{2}\right)$ and $\left(-\frac{1}{2}, \frac{9}{2}\right)$.

Helpful Hint: Do you try the Practice Problems before start-ing your homework? You should. If you worked Practice Prob-lem #1 in this section, you may have already learned to think about simplifying a function before starting to differentiate. Often there is no need to simplify first, but in Exercise 37 it really helps. Notice that

$$y = \frac{x^2+3x-1}{x} = x + 3 - \frac{1}{x},$$

so that

$$\frac{dy}{dx} = 1 + \frac{1}{x^2}.$$

43. Recall that the marginal revenue, MR, is defined by

$$MR = R'(x).$$

Now, the average revenue is maximized when

$$\frac{d}{dx}(AR) = 0.$$

To compute this derivative we use the quotient rule.

$$\frac{d}{dx}(AR) = \frac{d}{dx}\left(\frac{R(x)}{x}\right) = \frac{x \cdot R'(x) - R(x) \cdot (1)}{x^2}.$$

Set this derivative equal to zero. If $\dfrac{xR'(x) - R(x)}{x^2} = 0$

then the numerator must be zero.

$$xR'(x) - R(x) = 0,$$

$$xR'(x) = R(x),$$

$$R'(x) = \frac{R(x)}{x},$$

$$MR = AR.$$

49. We use the quotient rule to find the derivative of $\frac{f(x)}{1+x^2}$.

$$\frac{d}{dx}\left(\frac{f(x)}{1+x^2}\right) = \frac{(1+x^2)\cdot f'(x) - f(x)\cdot\frac{d}{dx}(1+x^2)}{(1+x^2)^2}$$

$$= \frac{(1+x^2)f'(x) - f(x)2x}{(1+x^2)^2}.$$

By hypothesis, $f'(x) = \frac{1}{1+x^2}$. Hence

$$\frac{d}{dx}\left(\frac{f(x)}{1+x^2}\right) = \frac{(1+x^2)\left(\frac{1}{1+x^2}\right) - f(x)\cdot 2x}{(1+x^2)^2} = \frac{1 - 2xf(x)}{(1+x^2)^2}.$$

Warning: Don't forget to use parentheses when appropriate in the product and quotient rules. Here is a common error:

$$\frac{d}{dx}\left(\frac{x^2 + x^3}{1 + e^x}\right) = \frac{1 + e^x(2x + 3x^2) - (x^2 + x^3)e^x}{(1 + e^x)^2}.$$

The numerator on the right should be

$$(1 + e^x)(2x + 3x^2) - (x^2 + x^3)e^x.$$

55. To find the derivative of $f(x)\, g(x)\, h(x)$, let

$$K(x) = g(x)\, h(x).$$

The derivative of $f(x)\, K(x)$ is:

$$f'(x)\, K(x) + f(x)\, K'(x) \quad \text{(by product rule)}$$

$$K'(x) = g'(x)\, h(x) + g(x)\, h'(x).$$

So,

$$(f(x)g(x)h(x))' = f'(x)\, \{g(x)h(x)\}$$

$$+ f(x)\, \{g'(x)h(x) + g(x)h'(x)\}$$

$$= f'(x)g(x)h(x) + f(x)g'(x)h(x) + f(x)g(x)h'(x).$$

3.2 The Chain Rule and the General Power Rule

The chain rule is used in subsequent chapters to derive formulas for the derivatives of composite functions where the outer function is either a logarithm function, an exponential function, or one of the trigonometric functions. In each case there will be a formula that has the same feel as the general power rule. These formulas will become so automatic that you will sometimes forget that they are special cases of the chain rule.

Many of the exercises in this section may be solved with the general power rule, but we approach the exercises with the abstract chain rule in mind. The practice you gain here will make it easier to use and remember the later versions of the chain rule.

1. If $f(x) = \frac{x}{x+1}$, $g(x) = x^3$, then $f(g(x)) = \frac{x^3}{x^3+1}$.

7. If $f(g(x)) = \sqrt{4-x^2}$, then $f(x) = \sqrt{x}$ and $g(x) = 4 - x^2$.

13. If $y = 6x^2(x-1)^3$, we have

$$\frac{dy}{dx} = \left[6x^2 \cdot \frac{d}{dx}(x-1)^3\right] + (x-1)^3 \cdot \frac{d}{dx}(6x^2) \quad \text{[Product Rule]}$$

$$= 6x^2 3(x-1)^2(1) + (x-1)^3 12x \quad \text{[Gen. Power Rule]}$$

We can factor $6x(x-1)^2$ out of each term to get

$$\frac{dy}{dx} = 6x(x-1)^2[3x+2(x-1)]$$

$$= 6x(x-1)^2(5x-2).$$

19. To differentiate $\left(\frac{4x-1}{3x+1}\right)^3$ let $f(x) = x^3$, $g(x) = \frac{4x-1}{3x+1}$, and use the chain rule. Now $f'(x) = 3x^2$ and

$$g'(x) = \frac{(3x+1)4 - (4x-1)3}{(3x+1)^2} \quad \text{[Quotient Rule]}$$

Hence

$$\frac{d}{dx}f(g(x)) = f'(g(x))g'(x)$$

$$= 3\left(\frac{4x-1}{3x+1}\right)^2 \cdot \frac{(3x+1)4 - (4x-1)3}{(3x+1)^2}.$$

This simplifies to

$$\frac{d}{dx}f(g(x)) = \frac{3(4x-1)^2(12x+4-12x+3)}{(3x+1)^4} = \frac{21(4x-1)^2}{(3x+1)^4}.$$

25. If $f(x) = x^5$ and $g(x) = 6x-1$, then $f'(x) = 5x^4$ and $g'(x) = 6$. Hence

$$\frac{d}{dx}f(g(x)) = f'(g(x))g'(x) = 5(6x-1)^4 \cdot 6.$$

31. If $f(x) = (x^3+1)^2$ and $g(x) = x^2+5$, then $f'(x) = 2(x^3+1)3x^2$ and $g'(x) = 2x$. Hence

$$\frac{d}{dx}f(g(x)) = f'(g(x))g'(x) = 2((x^2+5)^3+1)3(x^2+5)^2 2x$$

$$= 12x((x^2+5)^3+1)(x^2+5)^2.$$

37. If $y = u(u+1)^5$ and $u = x^2 + x$, then

$$\frac{dy}{du} = u5(u+1)^4 + (u+1)^5 = (u+1)^4[5u+(u+1)]$$

$$= (u+1)^4(6u+1)$$

$$\frac{du}{dx} = 2x + 1.$$

Hence $\frac{dy}{dx} = \frac{dy}{du} \cdot \frac{du}{dx} = (u+1)^4(6u+1)(2x+1)$. To express $\frac{dy}{dx}$ as a function of x alone, we substitute $x^2 + x$ for u to obtain

$$\frac{dy}{dx} = (x^2+x+1)^4(6x^2+6x+1)(2x+1).$$

43. A horizontal tangent line has slope 0; i.e., where $\frac{dy}{dx} = 0$.

$$\frac{dy}{dx} = 3(-x^2+4x-3)^2(-2x+4) \qquad \text{[Gen. Power Rule]}$$

Setting this equation equal to zero produces:

$$-2x+4 = 0, \qquad \text{or} \qquad -x^2+4x-3 = 0,$$
$$x = 2, \qquad\qquad\qquad x^2-4x+3 = 0,$$
$$\qquad\qquad\qquad\qquad\qquad (x-1)(x-3) = 0,$$
$$\qquad\qquad\qquad\qquad\qquad x = 1, \ x = 3.$$

Therefore, $x = 2, 1,$ and 3.

49. a. If $P = \dfrac{200x}{100+x^2}$, then the marginal profit is

$$\frac{dP}{dx} = \frac{(100+x^2)200 - 200x(2x)}{(100+x^2)^2} \qquad \text{[Quotient Rule]}$$

$$= \frac{200(100-x^2)}{(100+x^2)^2}.$$

b. If $x = 4 + 2t$, then $\dfrac{dx}{dt} = 2$. By the chain rule

$$\frac{dP}{dt} = \frac{dP}{dx}\cdot\frac{dx}{dt} = \frac{200(100-x^2)}{(100+x^2)^2}(2)$$

$$= \frac{400(100-x^2)}{(100+x^2)^2}.$$

To express $\dfrac{dP}{dt}$ as a function of t, we substitute 4 + 2t for x to obtain

$$\frac{dP}{dt} = \frac{400(100-(4+2t)^2)}{(100+(4+2t)^2)^2}.$$

c. When t = 8, $\dfrac{dP}{dt} = \dfrac{400(100-(4+16)^2)}{(100+(4+16)^2)^2} = -.48$.

Profits are falling at the rate of 480 dollars per week.

55. Since $\dfrac{d}{dx}f(g(x)) = f'(g(x))g'(x)$ we have $\dfrac{d}{dx}[f(g(x))]\big|_{x=1} = f'(g(1))g'(1)$. Since g(1) = 5, f'(5) = 4, and g'(1) = 6, we have f'(g(1))g'(1) = f'(5)6 = 4(6) = 24.

3.3 Implicit Differentiation and Related Rates

The material in this section is not used in the rest of the text. However, the two techniques introduced here are fundamental concepts of calculus and are commonly found in applications. Both techniques involve using the chain rule in situations where we don't know the specific formula for the inner part of a composite function.

1. We will differentiate $x^2 - y^2 = 1$ term by term. The first term x^2 has derivative 2x. We think of the second term y^2 as having the form $[g(x)]^2$. To differentiate we use the chain rule

$$\frac{d}{dx}[g(x)]^2 = 2[g(x)]g'(x).$$

Hence,

$$\frac{d}{dx}y^2 = 2y \cdot \frac{dy}{dx}.$$

On the right side of the original equation the derivative of the constant function 1 is zero. Thus implicit differentiation of $x^2 - y^2 = 1$ yields

$$2x - 2y \cdot \frac{dy}{dx} = 0.$$

Solving for $\frac{dy}{dx}$, we have

$$-2y \cdot \frac{dy}{dx} = -2x.$$

If $y \neq 0$,

$$\frac{dy}{dx} = \frac{x}{y}.$$

7. We will differentiate $2x^3 + y = 2y^3 + x$ term by term. The derivative at $2x^3$ is $6x$. We write $\frac{dy}{dx}$ for the derivative of y. On the right side of the equation, $2y^3$ has derivative $6y^2 \cdot \frac{dy}{dx}$ and x has derivative 1. Thus implicit differentiation of $2x^3 + y = 2y^3 + x$ yields

$$6x^2 + \frac{dy}{dx} = 6y^2 \cdot \frac{dy}{dx} + 1.$$

Solving for $\frac{dy}{dx}$ we have

$$(1-6y^2)\frac{dy}{dx} = 1 - 6x^2,$$

$$\frac{dy}{dx} = \frac{1 - 6x^2}{1 - 6y^2}.$$

13. To differentiate x^3y^2, use the product rule, treating y as a function of x:

$$\frac{d}{dx}x^3y^2 = x^3 \cdot \frac{d}{dx}(y^2) + y^2 \cdot \frac{d}{dx}(x^3)$$

$$= x^3 2y \cdot \frac{dy}{dx} + y^2(3x^2)$$

$$= 2x^3y\frac{dy}{dx} + 3x^2y^2.$$

3-11

Hence implicit differentiation of $x^3y^2 - 4x^2 = 1$ yields

$$2x^3y \frac{dy}{dx} + 3x^2y^2 - 8x = 0.$$

Solving for $\frac{dy}{dx}$ we have

$$2x^3y \frac{dy}{dx} = 8x - 3x^2y^2,$$

$$\frac{dy}{dx} = \frac{8x - 3x\ y}{2x^3 y} = \frac{8 - 3xy}{2x^2 y}.$$

19. Implicit differentiation of $4y^3 - x^2 = -5$ yields

$$12y^2 \cdot \frac{dy}{dx} - 2x = 0.$$

Solving for $\frac{dy}{dx}$ gives

$$12y^2 \frac{dy}{dx} = 2x,$$

$$\frac{dy}{dx} = \frac{2x}{12y^2}.$$

Hence,

$$\left. \frac{dy}{dx} \right|_{\substack{x=3 \\ y=1}} = \frac{2(3)}{12(1)^2} = \frac{1}{2}.$$

25. A two-step procedure is required.

(a) Find the slope at each point.
(b) Use the point-slope formula to obtain the equation of the tangent line.

For (a), note that

$$\frac{d}{dx}x^2y^4 = x^2 \cdot \frac{d}{dx}y^4 + y^4 \cdot \frac{d}{dx}x^2 = x^2 \cdot 4y^3 \frac{dy}{dx} + y^4 \cdot 2x.$$

Thus implicit differentiation of $x^2y^4 = 1$ yields

$$4y^3x^2 \frac{dy}{dx} + 2y^4x = 0,$$

$$4y^3x^2 \frac{dy}{dx} = -2y^4x,$$

$$\frac{dy}{dx} = \frac{-2y^4x}{4y^3x^2} = \frac{-y}{2x}.$$

At the point $\left(4,\frac{1}{2}\right)$, the slope of the curve is given by

$$\left.\frac{dy}{dx}\right|_{\substack{x=4 \\ y=1/2}} = \frac{-1/2}{2(4)} = -\frac{1}{16}.$$

Hence the equation of the tangent line at $\left(4,\frac{1}{2}\right)$ is

$$y - \frac{1}{2} = -\frac{1}{16}(x - 4).$$

At the point $\left(4,-\frac{1}{2}\right)$, the slope of the curve is given by

$$\left.\frac{dy}{dx}\right|_{\substack{x=4 \\ y=-1/2}} = \frac{-(-1/2)}{2(4)} = \frac{1}{16}.$$

Hence the equation of the tangent line at $\left(4,-\frac{1}{2}\right)$ is

$$y + \frac{1}{2} = \frac{1}{16}(x - 4).$$

31. We will differentiate $x^4 + y^4 = 1$ term by term. Since x is a function of t, the general power rule gives

$$\frac{d}{dt}x^4 = 4x^3 \cdot \frac{dx}{dt}.$$

Similarly,

$$\frac{d}{dt}y^4 = 4y^3 \cdot \frac{dy}{dt}.$$

And

$$\frac{d}{dt}(1) = 0.$$

Hence $4x^3 \frac{dx}{dt} + 4y^3 \frac{dy}{dt} = 0$. Solving for $\frac{dy}{dt}$ we have

$$4y^3 \frac{dy}{dt} = -4x^3 \frac{dx}{dt},$$

$$\frac{dy}{dt} = \frac{-x^3}{y^3} \frac{dx}{dt}.$$

37. First we compute $\frac{dy}{dt}$. Differentiating $x^2 - 4y^2 = 9$ term by term, we have

$$2x\frac{dx}{dt} - 8y\frac{dy}{dt} = 0.$$

Solving for $\frac{dy}{dt}$ we have

$$-8y\frac{dy}{dt} = -2x\frac{dx}{dt},$$

$$\frac{dy}{dt} = \frac{x}{4y}\frac{dx}{dt}.$$

The problem tells us that at the point $(5,-2)$, the x-coordinate is increasing at the rate of 3 units per second, that is, $\frac{dx}{dt} = 3$. Hence,

$$\frac{dy}{dt} = \frac{5}{4(-2)}(3) = -\frac{15}{8} \text{ units per second.}$$

43. $\frac{d}{dt}P^5V^7 = P^5 \cdot \frac{d}{dt}V^7 + V^7 \cdot \frac{d}{dt}P^5$

$$= P^5 7V^6\frac{dV}{dt} + V^7 5P^4\frac{dP}{dt}.$$

$\frac{d}{dx}(k) = 0$ (since k is a constant).

Hence $7P^5V^6\frac{dV}{dt} + 5P^4V^7\frac{dP}{dt} = 0$. Solving for $\frac{dV}{dt}$ we have

$$7P^5V^6\frac{dV}{dt} = -5P^4V^7\frac{dP}{dt},$$

$$\frac{dV}{dt} = \frac{-5P^4V^7}{7P^5V^6}\frac{dP}{dt},$$

$$= \frac{-5V}{7P}\frac{dP}{dt}.$$

The problem tells us that when $V = 4$ liters, $P = 200$ units and $\frac{dP}{dt} = 5$ units per second. Hence

$$\frac{dV}{dt} = \frac{-5(4)}{7(200)}(5) = -\frac{1}{14}.$$

Therefore the volume is decreasing at the rate of $\frac{1}{14}$ liters per second.

Chapter 3: Supplementary Exercises

The techniques of differentiation presented in the first two sections of this chapter must be mastered. They will be used throughout the text.

You may need to review curve sketching and optimization problems because these problems can be more difficult when they involve the product rule or quotient rule. On an exam that covers both Chapters 2 and 3 (or on the final exam) you may see a problem similar to Exercises 39 and 40 in Section 3.1 or Exercises 23 and 24 in Section 3.2.

1. $\frac{d}{dx}[(4x-1)(3x+1)^4] = (4x-1)\cdot\frac{d}{dx}(3x+1)^4 + (3x+1)^4\cdot\frac{d}{dx}(4x-1)$

$$\text{[Product Rule]}$$

$$= (4x-1)4(3x+1)^3\cdot 3 + (3x+1)^4\cdot 4.$$

If we factor $4(3x+1)^3$ from each term, we have

$$\frac{d}{dx}[(4x-1)(3x+1)^4] = 4(3x+1)^3[3(4x-1) + (3x+1)]$$

$$= 4(3x+1)^3[12x-3+3x+1]$$

$$= 4(3x+1)^3(15x-2).$$

7. If $y = 3(x^2-1)^3(x^2+1)^5$, then

$$\frac{dy}{dx} = 3(x^2-1)^3\cdot\frac{d}{dx}(x^2+1)^5 + (x^2+1)^5\cdot\frac{d}{dx}3(x^2-1)^3 \quad \text{[Prod. Rule]}$$

$$= 3(x^2-1)^3 5(x^2+1)^4 2x + (x^2+1)^5 9(x^2-1)^2 2x.$$

If we factor $6x(x^2-1)^2(x^2+1)^4$ from each term, we have

$$\frac{dy}{dx} = 6x(x^2-1)^2(x^2+1)^4(5(x^2-1)+3(x^2+1))$$

$$= 6x(x^2-1)^2(x^2+1)^4(8x^2-2)$$

$$= 12x(x^2-1)^2(x^2+1)^4(4x^2-1).$$

13. If $f(x) = (3x+1)^4(3-x)^5$, then

$$f'(x) = (3x+1)^4\cdot\frac{d}{dx}(3-x)^5 + (3-x)^5\cdot\frac{d}{dx}(3x+1)^4 \quad \text{[Prod. Rule]}$$

$$= (3x+1)^4 5(3-x)^4(-1) + (3-x)^5 4(3x+1)^3 3$$

$$= (3x+1)^3(3-x)^4[-5(3x+1) + 12(3-x)]$$

$$= (3x+1)^3(3-x)^4(-27x+31).$$

Set $f'(x) = 0$ and solve for x:

$$(3x+1)^3(3-x)^4(-27x+31) = 0.$$

$$x = -\frac{1}{3}, \; x = 3, \text{ and } x = \frac{31}{27}.$$

19. We want to evaluate $\frac{dC}{dt}$. The problem tells us that daily sales are rising at the rate of three lamps per day. This means $\frac{dx}{dt} = 3$. Now if $C = 40x + 30$, then $\frac{dC}{dx} = 40$. Hence by the chain rule,

$$\frac{dC}{dt} = \frac{dC}{dx} \cdot \frac{dx}{dt} = 40(3) = 120.$$

Costs are rising by \$120 per day.

25. $h(x) = f(g(x))$,

$h(1) = f(g(1))$, and since $g(1) = 2$, $h(1) = f(2) = 1$,

$$h'(x) = f'(g(x)) g'(x) \quad \text{[Chain rule]}.$$
$$h'(1) = f'(g(1)) g'(1)$$
$$= f'(2) g'(1).$$

The slope of the tangent line at $x = 2$ on the graph of $f(x)$ is -1. This is clear from an examination of the "graph paper" background in Fig. 2. Therefore, $f'(2) = -1$. Similarly, Fig. 2 shows that the slope of the tangent line at $x = 1$ on the graph of $g(x)$ is $\frac{3}{2}$, so

$$g'(1) = \frac{3}{2},$$

$$h'(1) = (-1)\frac{3}{2} = -\frac{3}{2}.$$

31. If $g(x) = \sqrt{x} = x^{1/2}$, then $g'(x) = \frac{1}{2}x^{-1/2}$. Since $f'(x) = x\sqrt{1-x^2}$, $\frac{d}{dx}f(g(x)) = f'(g(x))g'(x) = (x^{1/2})\sqrt{1-x}\left(\frac{1}{2}x^{-1/2}\right)$

$$= \frac{1}{2}\sqrt{1-x}.$$

37. If $u = \sqrt{x} = x^{1/2}$, then $\dfrac{du}{dx} = \dfrac{1}{2}x^{-1/2}$. Since $\dfrac{dy}{du} = \dfrac{u}{\sqrt{1+u^4}}$,

$$\frac{dy}{dx} = \frac{dy}{du}\cdot\frac{du}{dx} = \frac{u}{\sqrt{1+u^4}}\left(\frac{1}{2}x^{-1/2}\right).$$

To express $\dfrac{dy}{dx}$ as a function of x alone, we substitute $x^{1/2}$ for u to obtain

$$\frac{dy}{dx} = \frac{x^{1/2}}{(1+x^2)^{1/2}}\left(\frac{1}{2}x^{-1/2}\right) = \frac{1}{2\sqrt{1+x^2}}.$$

43. By the product rule,

$$\frac{d}{dx}x^2y^2 = x^2\cdot\frac{d}{dx}y^2 + y^2\cdot\frac{d}{dx}x^2 = x^2 2y\frac{dy}{dx} + y^2 2x.$$

Hence implicit differentiation of $x^2y^2 = 9$ yields

$$2x^2y\frac{dy}{dx} + 2xy^2 = 0.$$

Solving for $\dfrac{dy}{dx}$ we have

$$2x^2y\frac{dy}{dx} = -2xy^2$$

$$\frac{dy}{dx} = \frac{-2xy^2}{2x^2y} = -\frac{y}{x}.$$

If $x = 1$ and $y = 3$, then $\dfrac{dy}{dx} = -\dfrac{3}{1} = -3$.

49. First, to find $\dfrac{dx}{dt}$, we differentiate $6p + 5x + xp = 50$ term by term, with respect to t, using the produce rule on xp.

$$6\frac{dp}{dt} + 5\frac{dx}{dt} + x\frac{dp}{dt} + p\frac{dx}{dt} = 0.$$

Solving for $\dfrac{dx}{dt}$ we have

$$5\frac{dx}{dt} + p\frac{dx}{dt} = -6\frac{dp}{dt} - x\frac{dp}{dt},$$

$$(5+p)\frac{dx}{dt} = -(6+x)\frac{dp}{dt},$$

$$\frac{dx}{dt} = -\left(\frac{6+x}{5+p}\right)\frac{dp}{dt}.$$

If $x = 4$, $p = 3$, and $\frac{dp}{dt} = -2$, then

$$\frac{dx}{dt} = \left(\frac{6+4}{5+3}\right)2 = \frac{10}{4} = \frac{5}{2} = 2.5.$$

Therefore, the quantity is increasing at the rate of 2.5 units per unit time.

CHAPTER 4

THE EXPONENTIAL AND NATURAL LOGARITHM FUNCTIONS

4.1 Exponential Functions

The exercises in Section 4.1 focus on the specific skills you need later in Chapter 4. If you need more practice, go back to Section 0.5.

It will be helpful to make a table that lists the laws of exponents shown on page 288 and to add to it the "reverse" laws listed in the study guide notes for Section 0.5. (Use x and y in the exponents now instead of r and s.) You definitely need to memorize this enlarged list of laws. Working scores of problems will help, of course, because you will have to refer to your list. Here is an additional method. Write only the left side of each law in the table on page 288. Maybe mix up the order of the laws. On a separate sheet of paper, write the left side of the laws listed in the study guide for Section 0.5 (with x,y in place of r,s). Put these partial lists aside for a day or two while you work exercises from Section 4.1. Later, try to fill in the lists of laws without any help. You may need to do this more than once.

1. $4^x = (2^2)^x = 2^{2x}$,

   $(\sqrt{3})^x = (3^{1/2})^x = 3^{(1/2)x}$,

   $(\frac{1}{9})^x = (3^{-2})^x = 3^{-2x}$.

7. $2^{3x} \cdot 2^{-5x/2} = 2^{3x-5x/2} = 2^{6x/2-5x/2} = 2^{x/2} = 2^{(1/2)x}$

   $3^{2x} \cdot (\frac{1}{3})^{2x/3} = 3^{2x} \cdot (3^{-1})^{2x/3} = 3^{2x} \cdot 3^{-2x/3} = 3^{2x-2x/3}$

   $= 3^{(6x/3)-(2x/3)} = 3^{4x/3} = 3^{(4/3)x}$.

13. $6^x \cdot 3^{-x} = (2 \cdot 3)^x \cdot 3^{-x} = 2^x \cdot 3^x \cdot 3^{-x} = 2^x \cdot 3^0 = 2^x \cdot 1 = 2^x$,

$\dfrac{15^x}{5^x} = (\dfrac{15}{5})^x = 3^x$, or $\dfrac{15^x}{5^x} = (3 \cdot 5)^x \cdot 5^{-x} = 3^x \cdot 5^x \cdot 5^{-x} = 3^x$,

$\dfrac{12^x}{2^{2x}} = 12^x \cdot 2^{-2x} = (3 \cdot 2^2)^x \cdot 2^{-2x} = 3^x \cdot 2^{2x} \cdot 2^{-2x} = 3^x \cdot 2^0 = 3^x$,

or $\dfrac{12^x}{2^{2x}} = \dfrac{12^x}{4^x} = (\dfrac{12}{4})^x = 3^x$.

19. If $(2.5)^{2x+1} = (2.5)^5$, then equating exponents,

$$2x+1 = 5,$$
$$2x = 4,$$
$$x = 2.$$

25. If $(2^{x+1} \cdot 2^{-3})^2 = 2$, then

$$(2^{(x+1)-3})^2 = 2,$$
$$(2^{x-2})^2 = 2,$$
$$2^{2x-4} = 2^1.$$

Equating exponents, we have $2x - 4 = 1$ and hence $x = \dfrac{5}{2}$.

31. Since $2^{3+h} = 2^3(2^h)$, the missing factor is 2^h.

37. $3^{10x} - 1 = (3^{5x})^2 - 1 = (3^{5x}-1)(3^{5x}+1)$, since we know the more familiar result $x^2 - 1 = (x - 1)(x + 1)$. Therefore the missing factor is $3^{5x} + 1$.

4.2 The Exponential Function e^x

The purpose of this brief section and its exercises is to introduce the function e^x and to help you get used to working with it. Here are the main facts about e^x. The number e is between 2 and 3; the graph of the function e^x is similar to the graphs of 2^x and 3^x and lies between them. The number e is chosen so that the slope of $y = e^x$ is 1 at $x = 0$. The slope of $y = e^x$ at an arbitrary point (x, e^x) on the graph has the same numerical value as the y-coordinate of the point. That is (see Figure 3 on page 296),

$$[\text{slope at } (x, e^x)] = [\text{y-coord. of } (x, e^x)]$$

$$\frac{d}{dx} e^x = e^x.$$

1. If $y = 3^x$, then the slope of the secant line through $(0,1)$ and $(h, 3^h)$ is

$$\frac{3^h - 1}{h}.$$

As h approaches zero, the slope of the secant line approaches the slope of $y = 3^x$ at $x = 0$; i.e.,

$$\left. \frac{d}{dx} 3^x \right|_{x=0}.$$

If $h = .1$, then $\dfrac{3^h - 1}{h} = \dfrac{1.11612 - 1}{.1} = 1.1612.$

If $h = .01$, then $\dfrac{3^h - 1}{h} = \dfrac{1.01105 - 1}{.01} = 1.105.$

If $h = .001$, then $\dfrac{3^h - 1}{h} = \dfrac{1.00110 - 1}{.001} = 1.10.$

Therefore, $\left. \dfrac{d}{dx} 3^x \right|_{x=0} = \lim_{x \to 0} \dfrac{3^h - 1}{h} \approx 1.1.$

7. $\dfrac{d}{dx}(e^{4x}) = \dfrac{d}{dx}(e^x)^4 = 4(e^x)^3 \cdot \dfrac{d}{dx}(e^x)$ $\begin{bmatrix} \text{General} \\ \text{power rule} \end{bmatrix}$

$$= 4(e^x)^3 e^x = 4e^{3x} e^x = 4e^{4x}.$$

13. $\dfrac{1}{e^{-2x}} = e^{-(-2x)} = e^{2x}.$

19. If $e^{5x} = e^{20}$, then $5x = 20$, and $x = 4.$

25. By the quotient rule:

$$\frac{d}{dx}\left(\frac{e^x}{1 + e^x}\right) = \frac{(1+e^x) \cdot \frac{d}{dx}(e^x) - e^x \cdot \frac{d}{dx}(1+e^x)}{(1+e^x)^2}$$

$$= \frac{(1+e^x)e^x - e^x(e^x)}{(1+e^x)^2}$$

$$= \frac{e^x + e^{2x} - e^{2x}}{(1+e^x)^2}$$

$$= \frac{e^x}{(1+e^x)^2}.$$

4.3 Differentiation of Exponential Functions

Everything in this section is important. It is essential that you not fall behind the class at this point. Make every effort to completely finish the work in Sections 4.1–4.3 before the class begins Section 4.4. You may have difficulty with the material on logarithms in Sections 4.4–4.6 if you try to learn it while you are still uncertain about exponential functions. Be sure to read the boxes on pages 300–301. The importance of the differential equation $y' = ky$ will not be apparent until you reach Chapter 5, but it is desirable to think a little bit now about questions such as Exercises 45–48.

1. By formula (2) with $k = -1$,

$$\frac{d}{dx}(e^{-x}) = \frac{d}{dx}(e^{(-1)x}) = (-1)e^{(-1)x} = -e^{-x}.$$

7. If $f(x) = \frac{e^x - e^{-x}}{2} = \frac{1}{2}(e^x - e^{-x})$, then

$$f'(x) = \frac{1}{2}\left(\frac{d}{dx}e^x - \frac{d}{dx}e^{-x}\right) = \frac{1}{2}[e^x - (-e^{-x})], \text{ or } \frac{(e^x + e^{-x})}{2}.$$

13. $\frac{d}{dx}(\frac{1}{3}e^{3-2x}) = \frac{1}{3} \cdot \frac{d}{dx}(e^{3-2x})$

$$= \frac{1}{3}e^{3-2x} \cdot \frac{d}{dx}(3-2x) \qquad \left[\begin{array}{l}\text{Chain rule} \\ \text{for } e^{g(x)}\end{array}\right]$$

$$= \frac{1}{3}e^{3-2x}(-2) \qquad \text{[Don't forget the parentheses!]}$$

$$= -\frac{2}{3}e^{3-2x}.$$

19. If $f(x) = (2x+1-e^{2x+1})^4$, then

$$f'(x) = 4(2x+1-e^{2x+1})^3 \cdot \frac{d}{dx}(2x+1-e^{2x+1}) \qquad \left[\begin{array}{l}\text{General} \\ \text{power rule}\end{array}\right]$$

$$= 4(2x+1-e^{2x+1})^3(2-\frac{d}{dx}e^{2x+1})$$

$$= 4(2x+1-e^{2x+1})^3(2-e^{2x+1}(2)) \qquad \left[\begin{array}{l}\text{Chain rule} \\ \text{for } e^{g(x)}\end{array}\right]$$

$$= 4(2x+1-e^{2x+1})^3(2-2e^{2x+1}).$$

25. $\frac{d}{dx}(x+1)e^{-x+2} = (x+1) \cdot \frac{d}{dx}(e^{-x+2}) + e^{-x+2} \cdot \frac{d}{dx}(x+1) \qquad \left[\begin{array}{l}\text{Product} \\ \text{rule}\end{array}\right]$

$$= (x+1)(e^{-x+2}(-1)) + e^{-x+2}(1)$$

$$= e^{-x+2}(-x-1+1) = -xe^{-x+2}.$$

31. $f(x) = (1+x)e^{-x/2}$,

$$f'(x) = (1+x) \cdot \frac{d}{dx}e^{-x/2} + e^{-x/2} \cdot \frac{d}{dx}(1+x)$$

$$= (1+x)e^{-x/2} \cdot \left(-\frac{1}{2}\right) + e^{-x/2} \cdot 1$$

$$= e^{-x/2}\left(-\frac{1}{2} - \frac{x}{2} + 1\right)$$

$$= e^{-x/2}\left(\frac{1}{2} - \frac{x}{2}\right)$$

$$= \frac{1}{2}e^{-x/2}(1-x).$$

Set $f'(x) = 0$ and solve for x:

$$\frac{1}{2}e^{-x/2}(1-x) = 0.$$

Since $e^{-x/2}$ is never zero, we must have $1 - x = 0$. Hence $f'(x) = 0$ when $x = 1$.

Now if $f'(x) = \frac{1}{2}e^{-x/2}(1-x)$, then

$$f''(x) = \frac{1}{2}e^{-x/2} \cdot \frac{d}{dx}(1-x) + (1-x) \cdot \frac{d}{dx}\left(\frac{1}{2}e^{-x/2}\right)$$

$$= \frac{1}{2}e^{-x/2}(-1) + (1-x)\left(\frac{1}{2}\right)\left(-\frac{1}{2}\right)e^{-x/2}$$

$$= -\frac{1}{2}e^{-x/2}\left(1 + \frac{1}{2} - \frac{x}{2}\right)$$

$$= -\frac{1}{2}e^{-x/2}\left(\frac{3}{2} - \frac{x}{2}\right)$$

$$= -\frac{1}{4}e^{-x/2}(3-x).$$

Hence $f''(1) = -\frac{1}{4}e^{-1/2}(3-1) = -\frac{1}{4}e^{-1/2}(2)$.

This number is negative because e^x is always positive. Thus $f(x)$ is concave down at $x = 1$, and hence $x = 1$ is a relative maximum point.

37. If $f(x) = \frac{(x-1)^2}{e^x}$, then

$$f'(x) = \frac{e^x \cdot \frac{d}{dx}(x-1)^2 - (x-1)^2 \cdot \frac{d}{dx}(e^x)}{(e^x)^2}$$

$$= \frac{e^x \cdot 2(x-1) - (x-1)^2 e^x}{(e^x)^2} = \frac{e^x(x-1)[2-(x-1)]}{(e^x)^2}$$

$$= \frac{(x-1)(3-x)}{e^x}.$$

Clearly $f'(x) = 0$ when $x = 1$ or $x = 3$. To compute $f''(x)$, we write

$$f'(x) = \frac{4x - x^2 - 3}{e^x}.$$

$$f''(x) = \frac{e^x \cdot \frac{d}{dx}(4x-x^2-3) - (4x-x^2-3) \cdot \frac{d}{dx}(e^x)}{(e^x)^2}$$

$$= \frac{e^x(4-2x) - (4x-x^2-3)e^x}{(e^x)^2} = \frac{e^x[4-2x-(4x-x^2-3)]}{(e^x)^2}$$

$$= \frac{4 - 2x - 4x + x^2 + 3}{e^x} = \frac{x^2 - 6x + 7}{e^x}.$$

At $x = 1$, $f''(1) = \frac{1^2 - 6(1) + 7}{e^1} = \frac{2}{e} > 0$. The graph of $f(x)$ is concave up and $f(x)$ has a relative minimum at $x = 1$. At $x = 3$, $f''(3) = \frac{3^2 - 6(3) + 7}{e^3} = \frac{-2}{e^3} < 0$. So the graph of $f(x)$ is concave down and $f(x)$ has a relative maximum at $x = 3$.

43. $v(t) = 100,000e^{.2t}$, then

$$\frac{dv}{dt} = 100,000e^{.2t}(.2) = 20,000e^{.2t} \quad \text{(Rate of Appreciation).}$$

Now in 1995, $t = 5$, so

$$\left.\frac{dv}{dt}\right|_{t=5} = 20,000e^{.2(5)} = 20,000e \text{ dollars/year}$$

$$= 54,365.64 \text{ dollars/year.}$$

49. Result (3) says if $y' = ky$ (*), then

$$y = Ce^{kx}. \tag{*}$$

Let $y = f(x)$, and also let $g(x) = f(x) e^{-kx}$, differentiating the latter gives:

$$g'(x) = f'(x) e^{-kx} - kf(x)e^{-kx} = e^{-kx}[f'(x) - kf(x)]$$

$$= e^{-kx}[y'-ky] = 0 \text{ by } (*).$$

Thus $g'(x) = 0$, and therefore the only function whose derivative is zero for all x is the constant function, so $g(x) = C$. Remembering that $g(x) = f(x) e^{-kx}$, this implies that $f(x) e^{-kx} = C$, or $f(x) = Ce^{kx}$, and since $y = f(x)$:

$$y = Ce^{kx}.$$

4.4 The Natural Logarithm Function

The natural logarithm function is defined by its graph, and this is what you should think of when you are trying to understand what the ln x function really is. Study Fig. 2 in Section 4.4. The natural logarithm is a function whose graph is defined only for $x > 0$; the graph is increasing, passes through (1,0), and is concave downward. The graph of ln x is obtained by interchanging the x- and y-coordinates of points on the graph of $y = e^x$. This fact leads immediately to the fundamental relations (numbered below as in the text):

$$e^{\ln x} = x \qquad \text{(all positive x)} \qquad (2)$$

$$\ln e^x = x \qquad \text{(all x)} \qquad (3)$$

The main purpose of Section 4.4 is to teach you how to use these relations to solve equations involving ln x and e^x.

1. $\ln \dfrac{1}{e} = \ln e^{-1} = -1.$ [Relation (3)]

7. $\ln e^2 = 2.$ [Relation (3)]

13. If $e^{2x} = 5$, we take the (natural) logarithm of each side to obtain:

$$\ln e^{2x} = \ln 5,$$

$$2x = \ln 5, \qquad \text{[Relation (3)]}$$

$$x = \tfrac{1}{2} \ln 5.$$

19. The first step is to isolate the term $e^{-.00012x}$ so we can use the logarithm function to undo the effect of the exponentiation. Given $6e^{-.00012x} = 3$, we divide by 6 to obtain:

$$e^{-.00012x} = .5.$$

Taking the natural logarithm of each side, we have

$$\ln e^{-.00012x} = \ln .5,$$

$$-.00012x = \ln .5,$$

$$x = \frac{\ln .5}{-.00012}.$$

25. If $2e^{x/3} - 9 = 0$ then

$$2e^{x/3} = 9,$$

$$e^{x/3} = \frac{9}{2},$$

$$\ln(e^{x/3}) = \ln \frac{9}{2},$$

$$\frac{x}{3} = \ln \frac{9}{2}, \text{ or } x = 3 \ln \frac{9}{2}.$$

31. Given $4e^{x} \cdot e^{-2x} = 6$, observe that x occurs in two places. Use a property of exponents to write

$$4e^{-x} = 6.$$

Now proceed as in Exercise 19.

$$e^{-x} = \frac{6}{4} = \frac{3}{2},$$

$$\ln e^{-x} = \ln \frac{3}{2},$$

$$-x = \ln \frac{3}{2}, \text{ or } x = -\ln \frac{3}{2}.$$

Warning: A slight change in Exercise 31 can make it seem harder. Consider the equation

$$e^{x} = \frac{3}{2}e^{2x}.$$

Taking logarithms at this point is of no help because the equation $x = \ln(\frac{3}{2}e^{2x})$ expresses x in terms of something involving x. The correct step is to get x on only one side of the equation. There are two ways to do this:

(i) $e^{x} = \frac{3}{2}e^{2x}$,   or   (ii) $e^{x} = \frac{3}{2}e^{2x}$,

$e^{x} - \frac{3}{2}e^{2x} = 0$. $e^{x} \cdot e^{-2x} = \frac{3}{2}e^{2x} \cdot e^{-2x}$.

Equation (i) leads nowhere because you can't take the logarithm of each side, since $\ln 0$ is not defined. Also,

$\ln (e^x - \frac{3}{2}e^{2x})$ cannot be simplified. Equation (ii) is the correct approach, since a property of exponents leads to $e^{-x} = \frac{3}{2}e^0 = \frac{3}{2}$. This can be solved for x as in Exercise 31, yielding $x = -\ln \frac{3}{2}$.

37. If $f(t) = 5(e^{-.01t} - e^{-.51t})$, then

$$f'(t) = 5(e^{-.01t}(-.01) - e^{-.51t}(-.51))$$

$$= 5(-.01e^{-.01t} + .51e^{-.51t}).$$

Set $f'(t) = 0$ and solve for t:

$$5(-.01e^{-.01t} + .51e^{-.51t}) = 0,$$

$$-.01e^{-.01t} + .51e^{-.51t} = 0.$$

Thus,

$$-.01e^{-.01t} = -.51e^{-.51t},$$

$$e^{-.01t} \cdot e^{.51t} = \frac{-.51}{-.01} e^{-.51t} \cdot e^{.51t},$$

$$e^{.5t} = 51,$$

$$\ln e^{.5t} = \ln 51,$$

$$.5t = \ln 51, \text{ or } t = 2 \ln 51.$$

4.5 The Derivative of ln x

The exercises in this section combine the derivative formula for ln x with a review of the product rule, the quotient rule, and the chain rule. You should memorize the chain rule for the logarithm function:

$$\boxed{\frac{d}{dx}[\ln g(x)] = \frac{1}{g(x)} \cdot g'(x) = \frac{g'(x)}{g(x)}}$$

This rule is used in Exercises 1, 7, 13, and 19, discussed below.

1. $\dfrac{d}{dx}\ln 2x = \dfrac{1}{2x} \cdot \dfrac{d}{dx}(2x) = \dfrac{1}{2x}(2) = \dfrac{1}{x}.$

7. $\dfrac{d}{dx}e^{\ln x + x} = e^{\ln x + x} \cdot \dfrac{d}{dx}(\ln x + x) = \left(\dfrac{1}{x} + 1\right)e^{\ln x + x}.$

4-9

13. $\frac{d}{dx}\ln(kx) = \frac{1}{kx}\cdot\frac{d}{dx}(kx) = \frac{1}{kx}(k) = \frac{1}{x}$.

19. $\frac{d}{dx}\ln(e^{5x} + 1) = \frac{1}{e^{5x} + 1}\cdot\frac{d}{dx}(e^{5x} + 1)$

$$= \frac{1}{e^{5x} + 1}\cdot e^{5x}(5)$$

$$= \frac{5e^{5x}}{e^{5x} + 1}.$$

25. For the equation of the tangent line we need a point on the line and the slope of the line. We use the original equation $y = \ln(x^2+e)$ to find a point. If $x = 0$, then $y = \ln(0^2+e) = \ln e = 1$, and hence $(0,1)$ is on the line. For the slope of the tangent line, we first find the general slope formula.

$$\frac{dy}{dx} = \frac{1}{x^2 + e}\cdot\frac{d}{dx}(x^2 + e)$$

$$= \frac{1}{x^2 + e}(2x) \qquad [\text{e is a constant}]$$

$$= \frac{2x}{x^2 + e}.$$

The slope of the tangent line when $x = 0$ is $\frac{2(0)}{0^2 + e} = 0$.
Therefore the tangent line is the horizontal line passing through the point $(0,1)$. The equation of this line is $y - 1 = 0(x - 0)$; that is, $y = 1$.

31. The marginal cost at $x = 10$ is $C'(10)$. Using the quotient rule, we first find

$$C'(x) = \frac{d}{dx}\left(\frac{\ln x}{40-3x}\right)$$

$$= \frac{(40-3x)\cdot\frac{d}{dx}(\ln x) - (\ln x)\cdot\frac{d}{dx}(40-3x)}{(40-3x)^2}$$

$$= \frac{(40-3x)(\frac{1}{x}) - (\ln x)(-3)}{(40-3x)^2}.$$

Now $C'(10) = \dfrac{(40-3(10))\frac{1}{10} + 3 \ln 10}{(40-3(10))^2}$

$\qquad = \dfrac{10(\frac{1}{10}) + 3 \ln 10}{(10)^2} = \dfrac{1 + 3 \ln 10}{100}.$

4.6 Properties of the Natural Logarithm Function

Just as with laws of exponents, it is essential to know the properties of logarithms "backwards and forwards." You will find it helpful to add the following properties to the list on page 316.

LI′ $\quad \ln x + \ln y = \ln xy$
LII′ $\quad -\ln x = \ln(\frac{1}{x})$
LIII′ $\quad \ln x - \ln y = \ln(\frac{x}{y})$
LIV′ $\quad b \ln x = \ln(x^b)$

Unfortunately, many students make up additional "laws" that are not true. A study of the following facts will help you avoid the most common incorrect "laws".

a. $\ln(x + y)$ is <u>not</u> equal to $\ln x + \ln y$.

b. $\ln(e^x + e^y)$ is <u>not</u> equal to $x + y$.

c. $(\ln x)(\ln y)$ is <u>not</u> equal to $\ln x + \ln y$.

d. $\dfrac{\ln x}{\ln y}$ is <u>not</u> equal to $\ln x - \ln y$.

1. $\ln 5 + \ln x = \ln 5x$. $\hfill$ (LI′)

7. $e^{2\ln x} = e^{\ln x^2} = x^2$. $\hfill$ (LIV′)

13. We have $2 \ln 5 = \ln 5^2 = \ln 25$, $\hfill$ (LIV′)
    and $\quad 3 \ln 3 = \ln 3^3 = \ln 27$. $\hfill$ (LIV′)

   Since the graph of the natural logarithm function is increasing, and since $25 < 27$, we know that $\ln 25 < \ln 27$. Therefore, $3 \ln 3$ is larger.

Helpful Hint: Whenever you have a function to differentiate, you should pause and check if the form of the function can be simplified *before* you begin to differentiate. Exercises 19-24 illustrate how much this will simplify your work.

19. $y = \ln[(x+5)(2x-1)(4-x)] = \ln(x+5) + \ln(2x-1) + \ln(4-x)$.

$\qquad\qquad\qquad\qquad\qquad\qquad\qquad\qquad\qquad\qquad$ (LI′)

$\qquad \dfrac{dy}{dx} = \dfrac{d}{dx}\ln(x+5) + \dfrac{d}{dx}\ln(2x-1) + \dfrac{d}{dx}\ln(4-x)$

$\qquad\qquad = \dfrac{1}{x+5} + \dfrac{1.(2)}{2x-1} + \dfrac{1.(-1)}{4-x}$

$\qquad\qquad = \dfrac{1}{x+5} + \dfrac{2}{2x-1} + \dfrac{1}{x-4}.$

25. If $f(x) = (x+1)^3(4x-1)^2$

$\qquad\qquad \ln f(x) = \ln[(x+1)^3(4x-1)^2]$

$\qquad\qquad\qquad\quad = \ln(x+1)^3 + \ln(4x-1)^2$ $\qquad\qquad\qquad$ (LI′)

$\qquad\qquad\qquad\quad = 3\ln(x+1) + 2\ln(4x-1)$ $\qquad\qquad\quad$ (LIV′)

Differentiate both sides with respect to x:

$\qquad\qquad \dfrac{1}{f(x)}\cdot f'(x) = \dfrac{f'(x)}{f(x)} = \dfrac{3}{x+1} + \dfrac{2.}{4x-1}\cdot 4,$

so

$\qquad\qquad\qquad f'(x) = f(x)\left\{\dfrac{3}{x+1} + \dfrac{8}{4x-1}\right\}$

$\qquad\qquad\qquad\qquad = (x+1)^3(4x-1)^2\left\{\dfrac{3}{x+1} + \dfrac{8}{4x-1}\right\}.$

31. $f(x) = e^x\sqrt{x^2 - 1},$

$\qquad \ln f(x) = \ln e^x\sqrt{x^2 - 1}$

$\qquad\qquad\qquad = \ln e^x + \ln(x^2 - 1)^{\frac{1}{2}}$ $\qquad\qquad\qquad\qquad$ (LI′)

$\qquad\qquad\qquad = x\ln e + \dfrac{1}{2}\ln(x^2 - 1)$ $\qquad\qquad\qquad\qquad$ (LIV′)

$\qquad\qquad\qquad = x + \dfrac{1}{2}\ln(x^2 - 1).$

Differentiate both sides with respect to x:

$\qquad\qquad\qquad \dfrac{f'(x)}{f(x)} = 1 + \dfrac{1}{2}\left[\dfrac{1}{x^2-1}\right]2x$

$\qquad\qquad\qquad\qquad = 1 + \dfrac{x}{x^2-1},$

$$\text{so } f'(x) = f(x)\left[1 + \frac{x}{x^2-1}\right]$$

$$= e^x\sqrt{x^2-1}\left[1 + \frac{x}{x^2-1}\right].$$

37. $\ln y = k\ln x + \ln c,$

$e^{\ln y} = e^{k\ln x + \ln c} = e^{k\ln x}e^{\ln c}$ (by laws on
                                                                   p. 280 of text)

$$= e^{\ln x^k}e^{\ln c}. \hspace{3cm} \text{(LIV}')$$

Since $e^{\ln y} = y$, $e^{\ln x^k} = x^k$, and $e^{\ln c} = c$, the last equation becomes:

$$y = cx^k.$$

Review of Chapter 4

There are many facts in this chapter to remember and keep straight. The learning process requires time and lots of practice. Don't wait until the last minute to review. Your efforts to master this chapter will be rewarded later, since the exponential and natural logarithm functions appear in nearly every section in the rest of the text.

By now you should have constructed the expanded lists of properties of exponents and logarithms. To these lists add notes about common mistakes—yours and the ones mentioned in the study guide notes for this chapter. Here is another common error: If you take logarithms of each side of an equation of the form

$$A = B + C,$$

you *cannot* write $\ln A = \ln B + \ln C$. The correct form is

$$\ln A = \ln(B + C).$$

Similarly, if you exponentiate each side of $A = B + C$, you *cannot* write $e^A = e^B + e^C$. The correct form is

$$e^A = e^{(B + C)}.$$

Here are four problems that contain "traps" for the unwary student. Try them now. You will find answers later in the manual as you work through the supplementary exercises. *Please* don't look for the answers until you have done your best to work the problems.

(A) Solve for y in terms of x: $\ln y - \ln x^2 = \ln 5$.

(B) Solve for y in terms of x: $e^y - e^{-3} = e^{2x}$.

(C) Solve for x: $\dfrac{\ln 10x^3}{\ln 2x^2} = 1$.

(D) Simplify, if possible: $\ln(x^3 - x^2)$.

Chapter 4: Supplementary Exercises

In Exercises 9-14 and 35-40, *"simplify"* means to use the laws of exponents and logarithms to write the expression in another form that either is less complicated or at least is more useful for some purposes.

1. $27^{4/3} = (27^{1/3})^4 = 3^4 = 81$.

7. $\dfrac{9^{5/2}}{9^{3/2}} = 9^{5/2 - 3/2} = 9^{2/2} = 9$.

13. $(e^{8x} + 7e^{-2x})e^{3x} = e^{8x} \cdot e^{3x} + 7e^{-2x} \cdot e^{3x}$
$$= e^{11x} + 7e^x.$$

19. $\dfrac{d}{dx} 10e^{7x} = 10 \dfrac{d}{dx} e^{7x} = 10e^{7x}(7) = 70e^{7x}$.

25. $\dfrac{d}{dx}\left(\dfrac{x^2 - x + 5}{e^{3x} + 3}\right) = \dfrac{(e^{3x}+3) \cdot (2x-1) - (x^2-x+5) \cdot 3e^{3x}}{(e^{3x} + 3)^2}$ $\begin{bmatrix} \text{Quot.} \\ \text{rule} \end{bmatrix}$

Helpful Hint: In Exercises 27-30, the solution is a *function*, not a number. An equation in which both the unknown function and its (unknown) derivative appear is called a *differential equation*. The differential equations that appear here are solved with the boxed result (3) of Section 4.3.

31. $y = e^{-x} + x$, $y' = -e^{-x} + 1$, $y'' = e^{-x}$.

Now set $y' = 0$ and solve for x:
$$-e^{-x} + 1 = 0,$$
$$e^{-x} = 1,$$
$$\ln e^{-x} = \ln(1),$$
$$-x = 0, \text{ or } x = 0.$$

If $x = 0$, then $y = e^{-0} + 0 = 1$ and $y'' = e^{-0} = 1 > 0$, hence the graph is concave up at $(0,1)$. Thus the graph has a relative minimum at $(0,1)$. Since $y'' = e^{-x} > 0$ for all x, the graph is concave up for all x and there are no inflection points. As x becomes large: e^{-x} approaches 0, and for this case $e^{-x} + x$ is only slightly larger than x. Hence for large positive values of x, the graph of $y = e^{-x} + x$ has $y = x$ as an asymptote. The graph is shown in the answer section of the text.

Warning: Be careful with Exercise 35. Don't use an incorrect property of logarithms, and don't look at the answer until you have tried the problem!

37. $e^{-5\ln 1} = e^{-5(0)} = e^0 = 1.$

43. $2 \ln t = 5$, then
$$\ln t = \frac{5}{2},$$
$$e^{\ln t} = e^{5/2},$$
$$t = e^{5/2}.$$

49. $\dfrac{d}{dx}(\ln x)^2 = 2(\ln x) \cdot \dfrac{d}{dx}(\ln x) = 2(\ln x)\left(\dfrac{1}{x}\right) = \dfrac{2\ln x}{x}.$

55. $\ln(\ln\sqrt{x}) = \dfrac{1}{\ln\sqrt{x}} \cdot \dfrac{d}{dx}\ln\sqrt{x} = \dfrac{1}{\ln x^{1/2}} \cdot \dfrac{d}{dx}\ln x^{1/2}$

$$= \dfrac{1}{(1/2)\ln x} \cdot \dfrac{d}{dx}\left[\dfrac{1}{2}\ln x\right]$$

$$= \dfrac{2}{\ln x} \cdot \dfrac{1}{2} \cdot \dfrac{d}{dx}\ln x = \dfrac{1}{\ln x} \cdot \dfrac{1}{x} = \dfrac{1}{x\ln x}.$$

Helpful Hint: Always be alert to the possibility of simplifying a function before you differentiate it. This is particularly important when the function involves $\ln x$ or e^x. Observe that
$$\ln(\ln\sqrt{x}) = \ln(\ln x^{1/2}) = \ln\left(\dfrac{1}{2}\ln x\right) = \ln\dfrac{1}{2} + \ln(\ln x).$$

Thus
$$\dfrac{d}{dx}\ln(\ln\sqrt{x}) = \dfrac{d}{dx}\ln\dfrac{1}{2} + \dfrac{d}{dx}\ln(\ln x)$$

$$= 0 + \dfrac{1}{\ln x} \cdot \dfrac{d}{dx}\ln x = \dfrac{1}{\ln x} \cdot \dfrac{1}{x} = \dfrac{1}{x\ln x}.$$

Hints for extra problems:

(A) Get $\ln y$ by itself on the left, and use a property of logarithms to simplify $\ln 5 + \ln x^2$.

(B) Write $e^y = e^{2x} + e^{-3}$ and then take the logarithm of each side. Be careful!

(C) You made a mistake if you obtained any of the following equations from problem (C):

$$\frac{10x^3}{2x^2} = 1, \qquad \ln\left(\frac{10x^3}{2x^2}\right) = 1, \qquad \ln(10x^3 - 2x^2) = 1.$$

None of these equations is equivalent to the original equation. The correct first step is to multiply both sides of the equation in (C) by $\ln 2x^2$ and obtain $\ln 10x^3 = \ln 2x^2$. _Now_ can you solve for x?

(D) Is it true that $\ln(x^3 - x^2)$ is the same as $\dfrac{\ln x^3}{\ln x^2}$?

The correct solutions to (A) – (D) are at the end of this supplementary exercise set.

61. First take the natural logarithm of each side.

$$\ln f(x) = \ln[(x^2+5)^6(x^3+7)^8(x^4+9)^{10}]$$

$$= \ln(x^2+5)^6 + \ln(x^3+7)^8 + \ln(x^4+9)^{10} \qquad \text{(LI)}$$

$$= 6\cdot\ln(x^2+5) + 8\cdot\ln(x^3+7) + 10\cdot\ln(x^4+9). \qquad \text{(LIV)}$$

Now, take the derivative of each side and solve for $f'(x)$.

$$\frac{f'(x)}{f(x)} = 6\left[\frac{1}{x^2+5}\cdot 2x\right] + 8\left[\frac{1}{x^3+7}\cdot 3x^2\right] + 10\left[\frac{1}{x^4+9}\cdot 4x^3\right]$$

$$= \frac{12x}{x^2+5} + \frac{24x^2}{x^3+7} + \frac{40x^3}{x^4+9}.$$

$$f'(x) = f(x)\left[\frac{12x}{2} + \frac{24x^2}{3} + \frac{40x^3}{4}\right]$$

$$= (x^2+5)^6(x^3+7)^8(x^4+9)^{10}\left[\frac{12x}{x^2+5} + \frac{24x^2}{x^3+7} + \frac{40x^3}{x^4+9}\right].$$

67. $y = (\ln x)^2$,

$$y' = 2(\ln x) \cdot \frac{d}{dx}\ln x = \frac{2\ln x}{x}.$$

$$y'' = \frac{x \cdot \frac{d}{dx}(2\ln x) - (2\ln x)\frac{d}{dx}x}{x^2} = \frac{x \cdot 2\left(\frac{1}{x}\right) - 2\ln x}{x^2}$$

$$= \frac{2 - 2\ln x}{x^2}.$$

Set $y' = 0$ and solve for x:

$$\frac{2\ln x}{x} = 0.$$

A fraction is zero only when its numerator is zero. Thus $2\ln x = 0$, and $\ln x = 0$. Then $e^{\ln x} = e^0 = 1$, so $x = 1$. If $x = 1$, then $y = (\ln 1)^2 = 0$, and

$$y'' = \frac{2 - 2\ln 1}{1^2} = 2 > 0.$$

Hence, the curve is concave up at $(1,0)$. Now, set $y'' = 0$ and solve for x to find the possible inflection points.

$$\frac{2 - 2\ln x}{x^2} = 0,$$

$$2 - 2\ln x = 0,$$

$$-2\ln x = -2, \text{ or } \ln x = 1.$$

So,

$$e^{\ln x} = e^1, \text{ or } x = e.$$

If $x = e$, then $y = (\ln e)^2 = 1$, hence $(e,1)$ is the only possible inflection point. We have seen that the second derivative is positive at $x = 0$. When x is large, then $y = 2\ln x$ is negative, so the second derivative is negative. Thus the concavity must change somewhere, which shows that $(e,1)$ is the inflection point. The graph is shown in the answer section of the text.

Solutions to Extra Problems

(A) $\ln y - \ln x^2 = \ln 5$,

$\ln y = \ln 5 + \ln x^2$,

$\ln y = \ln 5x^2$.

Exponentiate each side to obtain $y = x^2$.

(B) $e^y = e^{2x} + e^{-3}$,

$\ln e^y = \ln(e^{2x} + e^{-3})$, $\left[\begin{array}{l}\text{Note: The right side}\\\text{cannot be simplified.}\end{array}\right]$

$y = \ln(e^{2x} + e^{-3})$.

(C) $\ln 10x^3 = \ln 2x^2$. Exponentiate to obtain

$10x^3 = 2x^2$.

Notice that x cannot be zero, since $\ln 10x^3$ is not defined when $x = 0$. So we may divide by $2x^2$ and obtain $5x = 1$ and hence $x = 1/5$.

(D) If you answered "yes" to the "Hint", then you have made one of the most common mistakes involving logarithms. The logarithm of a sum or difference *cannot be simplified*, unless that sum or difference can be written somehow as a product or quotient. For instance, you *may* write

$\ln(x^3 - x^2) = \ln[x^2(x - 1)]$ [now the expression inside the logarithm is a product]

$= \ln x^2 + \ln(x - 1)$

$= 2 \ln x + \ln(x - 1)$.

Whether or not this answer is "simplified" depends on the use to be made of the expression. See our remark at the beginning of the supplementary exercises.

CHAPTER 5

APPLICATIONS OF THE EXPONENTIAL AND NATURAL LOGARITHM FUNCTIONS

5.1 Exponential Growth and Decay

The first question to ask when considering an exponential growth and decay problem is, "Does the quantity increase or decrease with time?" For instance, populations increase and radioactive substances decrease. The formula for the quantity present after t units of time will have the form $P(t) = P_0 e^{kt}$ if increasing and $P(t) = P_0 e^{-\lambda t}$ if decreasing. The problems are primarily of the following types:

1. Given P_0 and k (or λ), find the value of P(t) for a specific time t. These problems are solved by just substituting the value of t into the formula.

2. Given P_0 and k (or λ), find the time when P(t) assumes a specific value, call it A. That is, solve P(t) = A for t. Problems of this type are solved by replacing P(t) by its formula, dividing by P_0 and using logarithms. Sometimes A is given as a multiple of P_0, in which case the P_0's cancel each other.

3. Given P_0 and the value of P(t) at some specific time t, solve for k (or λ). This problem reduces to one of the form "solve $P_0 e^{kt} = A$," where A is the value of P(t). This equation is solved by the method of 2 above.

4. A combination of the above types. The problem gives enough information to solve for k (or λ), but only asks for the value of P(t) at a future time or asks for the time at which P(t) attains a given size. The important point to realize here is that the problem really con-

sists of two parts. First, find k (or λ) and second, answer the question asked. Exercises 9 and 25 are of this type. Students usually have difficulty with such problems on exams since they try to answer the question posed immediately without doing the intermediate step.

1. Since $P(t)$ satisfies the differential equation $P'(t) = .07 \cdot P(t)$, we can conclude that $P(t) = P_0 e^{.07t}$, where P_0 is the initial population and .07 is the growth constant. Also, $P(0) = P_0 = 400$. Thus $P(t) = 400e^{.07t}$.

7. (a) The initial population is 5.4 billion on January 1, 1991, so $P(0) = P_0 = 5.4$. Since at any time population grows at a rate proportional to the population at that time, we conclude that $P(t) = P_0 e^{kt} = 5.4e^{kt}$. Since January 1, 1995, corresponds to $t = 4$,

$$P(4) = 5.4e^{k \cdot 4} = 6.0.$$

Thus,

$$e^{4k} = \frac{6}{5.4} = 1.111,$$

$$4k = \ln(1.111) \approx .10536,$$

$$k = \frac{.10536}{4} = .02634.$$

Therefore, $P(t) = 5.4e^{.02634t}$.

(b) January 1, 2010 is 19 years after January 1, 1991. Hence the population on January 1, 2010 is given by $P(19) = 5.4e^{(.02634)(19)} \approx 5.4(1.649) = 8.9$ billion people.

(c) We must find the time t at which $P(t) = 7$.

$$P(t) = 5.4e^{.02634t} = 7.$$

Thus,

$$e^{.02634t} = \frac{7}{5.4} \approx 1.3,$$

$$.02634t \approx \ln(1.3) \approx .262,$$

$$t = \frac{.262}{.02634} \approx 9.5 \text{ years after 1991.}$$

Hence the world's population will reach 7 billion in the year 2000.

13. This problem is essentially the first type of problem discussed in the introduction to this section of the study guide. P_0 and λ are given and we are asked for $P(10)$. The only twist is that λ is revealed by a differential equation.

(a) By the boxed result on page 326 of the text,

$$P(t) = Ce^{-.08t}, \text{ where } C = P(0).$$

Therefore, $P(t) = 30e^{-.08t}$.

(c) $P(10) = 30e^{-.08(10)} = 30e^{-.8} = 30(.44933) = 13.4799$,
$P(10) \approx 13.48$ grams.

19. Let the original amount be 1 (=100%). Since $\lambda = .00012$, the amount of C^{14} remaining after t years is $P(t) = 1 \cdot e^{-.00012t}$. The amount remaining after 4500 years is

$$P(4500) = e^{-.00012(4500)} = e^{-.54} = .58275.$$

Therefore, 58.275% remains.

25. This problem is the fourth type of problem discussed in the introduction to this section of the study guide. We are given enough information to solve for λ and are asked for the time at which $P(t)$ reaches a certain amount (1% of its original amount). The solution has two parts.

Part 1: Use the half-life to find λ.
$$P(t) = P_0 e^{-\lambda t},$$
$$P(28) = P_0 e^{-\lambda(28)} = .5P_0, \text{ (since after 28 years, } 1/2 \text{ remains)}$$
$$e^{-\lambda(28)} = .5,$$
$$-\lambda(28) = \ln .5 \approx -.69,$$
$$\lambda = \frac{-.69}{-28} \approx .025.$$

Part 2: Use λ to write the formula $P(t) = P_0 e^{-.025t}$, and use this formula to solve $P(t) = .01P_0$ for t. That is,

$$P_0 e^{-.025t} = .01P_0,$$
$$e^{-.025t} = .01, \text{ (we divided both sides of the equation by } P_0)$$
$$-.025t = \ln(.01) \approx -4.6,$$

$$t = \frac{-4.6}{-.025} = 184 \text{ years.}$$

31. To find T we first need to find the equation of the tangent line to this function at t = 0.

$$y = Ce^{-\lambda t}; \quad \frac{dy}{dt} = -\lambda\, Ce^{-\lambda t}.$$

Thus at (0,C);

$$\left.\frac{dy}{dt}\right|_{t=0} = -\lambda\, Ce^{-\lambda 0} = -\lambda C.$$

Therefore the equation of the tangent line is:

$$y - C = -\lambda C (t - 0),$$
$$y = C(1 - \lambda t).$$

Finally, since t = T when y = 0 we see that

$$0 = C(1 - \lambda T) \Rightarrow T = \frac{1}{\lambda}.$$

5.2 Compound Interest

The exercises in this section are similar to those in Section 5.1. A problem involving continuously compounded interest can be thought of as a problem about a population of money that is growing exponentially. The exercises in this section are of the same four types discussed in the notes for Section 5.1.

1. The initial amount is P = 1000 dollars. The interest rate per period (of one year) is i = .10. There are n = 2 interest periods in two years. So the compound amount is A = $1000(1 + .10)^2 = 1000(1.21) = 1210$ dollars.

7. Since interest is compounded 365 times a year, we can get a reasonable estimate of the amount in the account by assuming that interest is compounded continuously (this makes the calculation relatively simple). Thus

$$A = Pe^{rt} = 500e^{(.07)(3)} = 500e^{.21} \approx 500(1.23368)$$
$$= 616.84 \text{ dollars.}$$

The account contains approximately $616.84 after 3 years.

13. Using $A = Pe^{rt}$, we have $38,000 = 10,000e^{.15t}$. Thus

$$e^{.15t} = \frac{38,000}{10,000} = 3.8,$$

$$.15t = \ln 3.8 \approx 1.335,$$

$$t = \frac{1.335}{.15} = 8.9 \text{ years.}$$

19. After one year \$100 grows to A dollars, where

$$A = 100e^{(.07)(1)} = 100e^{.07} = 100(1.0725) = \$107.25.$$

Since after one year, \$100 earns \$7.25 in interest, and since \$7.25 is 7.25% of 100, the effective annual rate of interest is 7.25%.

25. Read through the answers and note that in some cases such as (b) and (d) a function is evaluated at a specified value of t, that is, the time t is known but the value of the function must be computed. In other cases such as (e) and (f) the time t is unknown and an equation is solved to find t. Next, note that some answers such as (b), (c), and (f) involve the function A(t), the balance in the account, while other answers such as (d) and (e) involve the derivative A′(t), the *rate* at which the balance in the account is changing. Finally, answers (a) and (g) relate to the specific type of function and differential equation that are associated with an investment whose interest is compounded continuously.

Now, before you read further in this solution, go back to the text, read each question, and try to find the appropriate answer. *After you have done this*, read the solu- tion that follows.

A. "How fast...balance...growing" relates to A′(t); "in 3 years" indicates that t = 3. Answer: (d), compute A′(3).

B. Answer: (a).

C. The question involves the amount A(t) in the account and its relation to the initial amount A(0); "how long" indicates that the time is unknown and must be found. Answer: (h), solve A(t) = 3A(0) for t.

D. The question involves the amount A(t), and the time is known to be t = 3. Answer: (b), compute A(3).

E. "When will" means that t must be found; "the balance" relates to A(t). Answer: (f), solve A(t) = 3 for t.

F. "When will" means that t must be found; "the balance... growing at the rate..." relates to A′(t). Answer: (e), solve A′(t) = 3 for t.

G. Answer: (c)   H. Answer: (g)

Helpful Hint: Exercise 25 is important because it tests your understanding of the basic concepts of exponential growth. Similar exercises are Exercise 11 of Section 5.1 and Exercise 19 of the Supplementary Exercises.

Helpful Hint: Use estimation to check that the answer to a growth or decay problem is reasonable.

(a) Suppose an investment doubles every eight years and you compute that it will increase tenfold in 20 years. Is this reasonable? Well, the investment will increase four-fold in 16 years and eightfold in 24 years. Therefore, the answer is not reasonable.

(b) Suppose $1000 is invested at 5% interest and you compute it will grow to about $1103 in two years. Is this reasonable? Well, the deposit will earn .05($1000) or $50 interest the first year and a little more than that (due to interest on the interest) during the second year. Therefore, it will earn a little more than $100 in interest in two years, and so the result is reasonable.

(c) Suppose that a radioactive material has a half-life of 5 years and you compute that 1/10 of the material will remain after 22 years. Is this reasonable? Consider the following table which was constructed solely from the fact that the material will halve every five years.

Number of years	5	10	15	20
Fraction of original remaining	$\frac{1}{2}$	$\frac{1}{4}$	$\frac{1}{8}$	$\frac{1}{16}$

The table shows that the correct answer lies between 15 and 20 years. Therefore, the answer of 22 is not reasonable.

(d) The *Rule of 70* says that an investment earning an interest rate of r % will double in about 70/r years. For instance, an investment earning 7 percent interest will double in about 70/7 or 10 years. Similarly, an investment doubling in d years has earned an interest rate of about 70/d percent per year. (In former times, bankers used a *rule of 72*, because they relied on mental arithmetic, and 72 is easily divisible by many common interest rates.)

(e) The *Rule of 70* can also be applied to exponential decay. A quantity with decay constant λ halves in about $70/(100\lambda)$ units of time. For instance, if the decay constant of a radioactive material is .5, with time measured in years, then the half-life of the material is about $70/(100 \cdot .5) = 70/50 = 1.6$ years. The following table was constructed with this estimation technique.

Decay constant	.1	.01	.001	.0001
Half-life	7	70	700	7000

5.3 Applications of the Natural Logarithm Function to Economics

The material in this section is not used in any other part of the book. Elasticity is one of the most important concepts of economics and provides insights into the pricing of goods and services. In addition to being able to compute the elasticity of demand, you should be able to interpret what it means for the demand to be, say, elastic. The box on page 352 summarizes this. Here is another more informal way to remember it. When demand is elastic, an increase in price causes such a decline in sales that the total revenue falls. When demand is inelastic, an increase in price causes only a relatively small decline in sales, so that the total revenue still increases.

1. $f(t) = t^2$, so $f'(t) = 2t$.

   Thus at $t = 10$, $\dfrac{f'(10)}{f(10)} = \dfrac{2(10)}{10^2} = \dfrac{20}{100} = .2 = 20\%$.

   At $t = 50$, $\dfrac{f'(50)}{f(50)} = \dfrac{2(50)}{50^2} = \dfrac{100}{2500} = .04 = 4\%$.

7. $f(p) = \dfrac{1}{p + 2}$, thus $f'(p) = \dfrac{-1}{(p + 2)^2}$.

   So at $p = 2$, $\dfrac{f'(2)}{f(2)} = \dfrac{-\frac{1}{16}}{\frac{1}{4}} = -\dfrac{4}{16} = -.25 = -25\%$.

   And at $p = 8$, $\dfrac{f'(8)}{f(8)} = \dfrac{-\frac{1}{100}}{\frac{1}{10}} = -\dfrac{10}{100} = -.1 = -10\%$.

13. $f(p) = q = 700 - 5p$, so $f'(p) = -5$.

   $$E(p) = \frac{-pf'(p)}{f(p)} = \frac{-p(-5)}{700 - 5p} = \frac{5p}{700 - 5p} = \frac{p}{140 - p}.$$

   So at $p = 80$, $E(80) = \dfrac{80}{140 - 80} = \dfrac{80}{60} = \dfrac{4}{3}$.

   Since $\frac{4}{3} > 1$, demand is elastic.

19. (a) $f(p) = q = 600(5 - \sqrt{p}) = 3000 - 600p^{1/2}$, and thus

   $f'(p) = -300p^{-1/2}$.

   $$E(p) = \frac{-pf'(p)}{f(p)} = \frac{-p(-300p^{-1/2})}{3000 - 600p^{1/2}} = \frac{300p^{1/2}}{3000 - 600p^{1/2}}$$

   $$= \frac{p^{1/2}}{10 - 2p^{1/2}}.$$

Thus at p = 4, $E(4) = \dfrac{4^{1/2}}{10 - 2(4)^{1/2}} = \dfrac{2}{10 - 2(2)} = \dfrac{1}{3}$.

Since $\dfrac{1}{3} < 1$, demand is inelastic.

(b) Since demand is inelastic, an increase in price will bring about an increase in revenue. Thus, the price of a ticket should be raised.

25. $E_c(x) = \dfrac{\dfrac{d}{dx}\ln C(x)}{\dfrac{d}{dx}\ln x} = \dfrac{\dfrac{C'(x)}{C(x)}}{\dfrac{1}{x}} = \dfrac{xC'(x)}{C(x)}$.

5.4 Further Exponential Models

This section illustrates the wide variety of applications in which exponential functions appear. The material in this section is not needed for any other part of the book. The logistic curve will be studied further in Chapter 10. However, the discussion there is completely independent of the discussion in Section 5.4.

1. (a) $f(x) = 5(1 - e^{-2x}) = 5 - 5e^{-2x}$, $x \geq 0$. Thus $f'(x) = 10e^{-2x}$. Since $e^{-2x} > 0$ for every value of x, $10e^{-2x} > 0$ for every value of x, in particular when $x \geq 0$. Thus f(x) is increasing. $f''(x) = -20e^{-2x}$. Again, since $e^{-2x} > 0$, we have $-20e^{-2x} < 0$ for all values of x, in particular when $x \geq 0$. Thus f(x) is concave down.

   (b) $f(x) = 5(1 - e^{-2x})$, $x \geq 0$.
   Note that as x gets larger, $e^{-2x} = \dfrac{1}{e^{2x}}$ gets closer and closer to zero. Hence when x is very large the values of f(x) are very close to 5. (The values of f(x) are slightly less than 5 because $1 - e^{-2x}$ is slightly less than 1.)

   (c) See the graph in the answer section of the text.

7. The number of people who have heard about the indictment by time t is given by $f(t) = P(1 - e^{-kt})$, where P is the total population. Since after one hour, one quarter of the citizens had heard the news, we have
$f(1) = P(1 - e^{-k(1)}) = \dfrac{1}{4}P = .25P$. Hence

$$1 - e^{-k} = .25,$$

$$e^{-k} = .75,$$

$$-k = \ln .75 \approx -.29, \text{ or } k = .29.$$

Before continuing with the solution, write out the formula for f(t) for future reference.

$$f(t) = P(1 - e^{-.29t}).$$

Now we must find time t when $f(t) = \frac{3}{4}P = .75P$, that is,

$$P(1 - e^{-.29t}) = .75P.$$

Thus,

$$1 - e^{-.29t} = .75,$$

$$e^{-.29t} = .25,$$

$$-.29t = \ln .25 \approx -1.4,$$

$$t = \frac{-1.4}{-.29} = \frac{1.4}{.29} \approx 4.8 \text{ hours.}$$

13. (a) After 1 week, the length is f(1) = 2 cm.

(b) The rate of growth after 10 weeks is f'(10) = 2 cm/wk.

(c) We must solve f(t) = 10 for t. The horizontal line y = 10 crosses the graph of y = f(t) at the point (5, 10). Therefore, f(5) = 10 and the weed is 10 centimeters long after 5 weeks.

(d) We must solve f'(t) = 2 for t. The horizontal line y = 2 crosses the graph of y = f'(t) at t = 3 and t = 10. Therefore, the weed is growing at the rate of 2 centimeters per day after 3 weeks and after 10 weeks.

(e) The weed is growing at the fastest rate when f(t) has an inflection point, that is, when f"(t) is zero. This occurs at t = 6.4. Therefore, the weed is growing fastest after 6.4 weeks. At that time, the length of the weed is f(6.4) = 15 cm.

Chapter 5: Supplementary Exercises

1. P'(x) = -.2P(x) implies that P(x) is an exponential function of the form $P(x) = P_0 e^{-.2x}$, where $P_0 = P(0)$ is the atmospheric pressure at sea level. Thus $P_0 = 29.92$, and $P(x) = 29.92 e^{-.2x}$.

7. (a) The initial population (at t = 0) is 14.2 million. At time t = 10, the population is 17 million. Thus

$$P(t) = 14.2e^{kt}, \text{ and } P(10) = 14.2e^{k \cdot 10} \approx 17.$$

Solving for k,

$$e^{10k} = \frac{17}{14.2} \approx 1.197,$$

$$10k \approx \ln 1.197 \approx .1798,$$

$$k = \frac{.1798}{10} = .01798.$$

Thus $P(t) = 14.2e^{.01798t}$.

(b) For the year 2000, t = 20. Hence

$$P(20) = 14.2e^{(.01798)(20)}$$

$$= 14.2e^{.3596}$$

$$= 14.2(1.4328) = 20.346 \text{ million people.}$$

(c) We must find the time t when P(t) = 25. Hence

$$14.2e^{.01798t} = 25,$$

$$e^{.01798t} = \frac{25}{14.2} = 1.76,$$

$$.01798t = \ln 1.76 \approx 0.5653,$$

$$t = \frac{.5653}{.01798} = 31.1 \text{ years after 1980.}$$

Thus the population will reach 25 million in 2011.

13. E(8) = 1.5. So when the price is set at 8 dollars, a small increase in price will result in a relative rate of decrease in quantity demanded of about 1.5 times the relative rate of increase of price. Since a price increase of $.16 represents a 2% increase in price, the quantity demanded will decrease by 1.5(2%) = 3%. Since E(8) = 1.5 > 1, the demand is elastic. Hence the price increase yields a decrease in revenue.

19. First, read the answers given in the problem statement. Review the solution of Exercise 25 in Section 5.2. Then go back to the text, read each question and try to find the appropriate answer. *After you have done this,* read the solutions that follow, on the next page.

19. A. Answer: (g), $y' = ky$.

   B. "How fast" indicates that the rate $P'(t)$ is involved; "in 1/2 year" means that $t = .5$.
   Answer: (d), compute $P'(.5)$.

   C. Answer: (h), $P_0 e^{kt}$, $k < 0$.

   D. "Half-life" is the time t such that the amount $P(t)$ is one-half of the initial amount $P(0)$.
   Answer: (a), solve $P(t) = .5P(0)$ for t.

   E. "How many grams" relates to $P(t)$; "after 1/2 year" means that $t = .5$.
   Answer: (c), compute $P(.5)$.

   F. "When will" means that t must be found; "at the rate of 1/2 gram per year" relates to $P'(t)$.
   Answer: (f), solve $P'(t) = .5$ for t.

   G. "When will" means that t must be found; "1/2 gram remaining" involves $P(t)$.
   Answer: (b), solve $P(t) = .5$ for t.

   H. "How much" relates to $P(t)$. "Initially" refers to the time $t = 0$.
   Answer: (e), compute $P(0)$.

CHAPTER 6

THE DEFINITE
INTEGRAL

6.1 Antidifferentiation

The problems in this section remind us of Johnny Carson's routine where he gives the answer to a question and then determines the question. (For instance, in one case the answer was 9W and the question was "Do you spell your name with a V, Mr. Wagner?") In this section the answer is the derivative of a function and the question is "What is the function?"

To find an antiderivative, first make an educated guess and then differentiate the guess. (This may be done mentally if the differentiation is simple.) Your guess should be based on your experience with derivatives. In most cases, functions and their antiderivatives are the same basic type, such as power functions or exponential functions. If your first guess for the antiderivative is correct, you are finished (after you add "+C"). With some practice, your guess usually will be correct except for a constant factor, and you will only need to multiply the guess by a suitable constant. Exercises 7 to 22 show the correct form of the antiderivative. All you have to do is to adjust the constant k correctly.

Exercises 1-6 and 23-28 simply require antidifferentiation. The answers all contain "+C"; that is, each answer is an infinite collection of functions. In Exercises 33-36 and 47-53, each answer is a single function. This function is found by first antidifferentiating and then using additional information to find the proper value of C. On an exam or homework assignment, this value of C should be substituted for C in the antiderivative to obtain the answer.

1. Since $\frac{d}{dx}x^2 = 2x$, we see that $\frac{d}{dx}\frac{1}{2}x^2 = x$. Hence one anti-

    derivative of f(x) = x is F(x) = $\frac{1}{2}x^2$, and all antideriva-

    tives are of the form F(x) = $\frac{1}{2}x^2$ + C.

7. Since $\int x^{-5}\,dx = kx^{-4} + C$, we must have

$$\frac{d}{dx}(kx^{-4}+C) = x^{-5},$$

$$-4kx^{-5} = 1 \cdot x^{-5},$$

$$-4k = 1, \text{ or } k = -1/4.$$

13. Since $\int 5e^{-2t}\,dt = ke^{-2t} + C$, we must have

$$\frac{d}{dt}(ke^{-2t}+C) = 5e^{-2t},$$

$$-2ke^{-2t} = 5e^{-2t},$$

$$-2k = 5, \text{ or } k = -5/2.$$

19. As before, we must find the value of k for which

$$\frac{d}{dx}(k\ln|x+4|+C) = (x+4)^{-1},$$

$$k \cdot \frac{1}{x+4} = 1 \cdot \frac{1}{x+4}, \text{ so } k = 1.$$

Helpful Hint: The antiderivatives in Exercises 15, 16, 21, and 22 are a little harder to find because you must recall how the chain rule works. (Later we will study a general method for handling such problems. For now, however, you may use the technique of simply guessing the general form of an antiderivative.) For instance,

$$\frac{d}{dx}(3x+2)^5 = 5(3x+2)^4 \cdot \frac{d}{dx}(3x+2)$$

$$= 5(3x+2)^4 \cdot 3.$$

In Exercise 21, you must find k such that

$$\frac{d}{dx}k(3x+2)^5 = (3x+2)^4.$$

Remember that you can check your final answer by differentiating what you think is the correct antiderivative.

25. Rewriting the integral, we have

$$\int\left(\frac{2}{\sqrt{x}} - 3\sqrt{x}\right) dx = \int(2x^{-1/2} - 3x^{1/2})\,dx.$$

Since $\frac{d}{dx}4x^{1/2} = 2x^{-1/2}$, and $\frac{d}{dx}2x^{3/2} = 3x^{1/2}$, we have

$$\int(2x^{-1/2} - 3x^{1/2})\,dx = 4x^{1/2} - 2x^{3/2} + C$$

$$= 4\sqrt{x} - 2x^{3/2} + C.$$

31. From Theorem II, we see that if $f'(t) = 0$ for all t, then $f(t) = C$ for some constant C.

37. $f(x)$ is an antiderivative of $\frac{2}{x}$. Since $\frac{d}{dx} \ln x = \frac{1}{x}$, we have $\frac{d}{dx}(2 \ln x) = \frac{2}{x}$. So $2 \ln x$ is an antiderivative of $\frac{2}{x}$, and $f(x) = 2 \ln x + C$ for some C. To make the graph of $f(x)$ pass through $(1,2)$, we want $f(1) = 2$. That is,

$$2 \ln 1 + C = 2.$$

Since $\ln 1 = 0$, we find $C = 2$. Thus $f(x) = 2 \ln x + 2$.

43. $g(x) = f(x) + 3$,

$g'(x) = f'(x)$,

$g'(5) = f'(5) = \frac{1}{4}$.

49. We let $f(t)$ represent the temperature in degrees Celsius of the strawberries at time t. Then $f'(t) = 10e^{-.4t}$, and $f(0) = -5$ (since the initial temperature of the strawberries is $-5°C$). Thus $f(t) = \int 10e^{-.4t} \, dt = -25e^{-.4t} + C$. To find C, we use the fact that $f(0) = -25 + C = -5$. Hence $C = 20$, and $f(t) = -25e^{-.4t} + 20 = 20 - 25e^{-.4t}$.

55. $C'(x) = 1000 + 50x$.
Find the antiderivative, $F(x)$, of this expression. Noting that

$$\frac{d}{dx}(1000x) = 1000,$$

and

$$\frac{d}{dx}(x^2) = 2x, \text{ so } \frac{d}{dx}(25x^2) = 50x.$$

Therefore, $\frac{d}{dx}(1000x + 25x^2) = 1000 + 50x$, giving us the antiderivative as $F(x) = 1000x + 25x^2 + K$ (K constant). But $C(x) = F(x) = 1000x + 25x^2 + K$, and with the condition $C(0) = 10,000$, we can evaluate K: $K = 10,000$. Therefore, finally the cost function is:

$$C(X) = 25x^2 + 1000x + 10,000.$$

6.2 Areas and Riemann Sums

Riemann (pronounced "Reemahn") sums are introduced here in order to approximate the area under the graph of a nonnegative function, but their importance extends far beyond that. As we shall see later, Riemann sums are used to derive some important formulas in applications.

Computations with Riemann sums can be time-consuming, but some numerical homework is necessary for a proper understanding. Check with your instructor about how far to carry calculations on an exam. In some cases you may only have to write down the complete Riemann sum (using numbers and not symbols such as Δx), and omit the final arithmetic computation.

One goal of this section is to suggest how the change in a function over some interval is related to the area under the graph of the derivative of that function. See Example 4 and the Table on page 392. This concept will be made precise in Section 6.3.

1. $a = 0$, $b = 2$, $n = 4$, so $\Delta x = \dfrac{b-a}{n} = \dfrac{2-0}{4} = .5$.

   The first midpoint is $a + \Delta x/2 = 0 + .5/2 = .25$. Subsequent midpoints are found by adding Δx repeatedly. So the other midpoints are:

   $$.25 + .5 = .75,$$

   $$.75 + .5 = 1.25,$$

   $$1.25 + .5 = 1.75.$$

Helpful Hint: You may use fractions, if you wish, and write 1/4 in place of .25, and so on. However, you can use a calculator more efficiently if you convert to decimals.

Helpful Hint: To draw a partition that has 5 subintervals, draw a horizontal line and, beginning at one end, mark off five equal subintervals. This is easier than trying to divide one larger interval into five equal pieces. (See Exercises 3, 4, and 7-10.)

7. $\Delta x = (b - a)/n = (3 - 1)/5 = 2/5 = .4$ The first left endpoint is $x_1 = a = 1$. Subsequent left endpoints are found by adding Δx repeatedly.

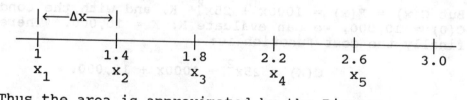

Thus the area is approximated by the Riemann sum:

$$\Delta x \, [f(1) + f(1.4) + f(1.8) + f(2.2) + f(2.6)]$$

$$= (.4.)[(1)^3 + (1.4)^3 + (1.8)^3 + (2.2)^3 + 2.6)^3]$$

$$= (.4)[1 + 2.744 + 5.832 + 10.648 + 17.576]$$

$$= (.4)[37.8] = 15.12.$$

13. $\Delta x = [1 - (-1)]/5 = 2/5 = .4.$ The first midpoint is:

$$x_1 = a + \Delta x/2 = -1 + .4/2 = -1 + .2 = -.8.$$

The successive midpoints are found by adding Δx to each of the previous midpoint values:

$$x_2 = -.8 + .4 = -.4, \quad x_3 = -.4 + .4 = 0,$$
$$x_4 = 0 + .4 = .4, \text{ and } x_5 = .4 + .4 = .8.$$

The area is approximated by the Riemann sum:

$$\Delta x \, [f(-.8) + f(-.4) + f(0) + f(.4) + f(.8)]$$

$$= (.4)\left\{ \sqrt{1 - (-.8)^2} + \sqrt{1 - (-.4)^2} + \sqrt{1 - (0)^2} \right.$$

$$\left. + \sqrt{1 - (.4)^2} + \sqrt{1 - (.8)^2} \right\}$$

$$= (.4)\left\{ \sqrt{.36} + \sqrt{.84} + 1 + \sqrt{.84} + \sqrt{.36} \right\}$$

$$= (.4)\{.6 + .916515 + 1 + .916515 + .6\}$$

$$= (.4)\{4.03303\} = 1.61321.$$

The error is $1.61321 - 1.57080 = 0.04241$ to 5 dec. places.

19. (a) This represents the increase in population from 1910 to 1950 in millions.

(b) The rate of cigarette consumption t years after 1985.

(c) 20 to 50.

25. The relationship is described as:

Area under the graph of $f(x)$ + Area under graph of $g(x)$

$$= \text{Area under the graph of } f(x) + g(x).$$

Proof: The area under $f(x)$ is approximately:

$$A_1 = \Delta x \{f(x_1) + f(x_2) + \ldots + f(x_n)\},$$

where x_1, x_2, ..., are the midpoints, the area under $g(x)$ is approximately

$$A_2 = \Delta x \{g(x_1) + g(x_2) + \ldots + g(x_n)\},$$

adding A_1 to A_2 produces,

$$\Delta x \{[f(x_1)+g(x_1)]+[f(x_2)+g(x_2)]+ \ldots +[f(x_n)+g(x_n)]\},$$

which is exactly the midpoint sum under the graph of $f(x)$ + $g(x)$. The initial statement is proven.

6.3 Definite Integrals and the Fundamental Theorem

Although definite integrals are defined by Riemann sums, they are usually calculated as the net change in an antiderivative. Since you have computed a few Riemann sums, you probably appreciate the value of the Fundamental Theorem of Calculus.

1. $\displaystyle\int_{-1}^{1} x\, dx = \frac{1}{2}x^2 \Big|_{-1}^{1} = \frac{1}{2} - \frac{1}{2} = 0.$

7. $\displaystyle\int_{0}^{1} 4e^{-3x}\, dx = -\frac{4}{3}e^{-3x}\Big|_{0}^{1} = [-\frac{4}{3}e^{-3}] - [-\frac{4}{3}]$

$$= \frac{4}{3} - \frac{4}{3}e^{-3}, \text{ or } \frac{4}{3}(1 - e^{-3}).$$

13. $\displaystyle\int_{3}^{6} x^{-1}\, dx = \ln|x| \Big|_{3}^{6} = \ln 6 - \ln 3 = \ln\frac{6}{3} = \ln 2.$

19. $\displaystyle\int_{0}^{3} (x^3+x-7)\, dx = (\frac{1}{4}x^4 + \frac{1}{2}x^2 - 7x)\Big|_{0}^{3} = [\frac{81}{4} + \frac{9}{2} - 21] - [0]$

$$= \frac{81}{4} + \frac{18}{4} - \frac{84}{4} = \frac{15}{4} = 3\frac{3}{4}.$$

25. The desired area is

$$\int_{0}^{1} e^{x/2}\, dx = 2e^{x/2}\Big|_{0}^{1} = 2e^{1/2} - 2 = 2(e^{1/2}-1).$$

31. The desired area is

$$\int_{0}^{3} 2x\, dx = x^2 \Big|_{0}^{3} = 9 - 0 = 9.$$

To check the answer with elementary geometry, consider the graph of the function $y = 2x$ at the right. Using the fact that the area of a triangle is given by [Area] $= \frac{1}{2}$[base][height], we see that the base has length 3, and the height is 6. Hence the area is $\frac{1}{2}(3)(6) = 9$.

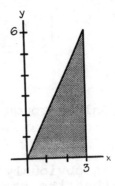

37. Number of cigarettes sold from 1980 to 1990 is

$$= \int_{20}^{30} (.1t + 2.4)\, dt$$

$$= \left. \frac{.1t^2}{2} + 2.4t \right|_{20}^{30} \quad \left(\frac{.1}{2} = .05 \right)$$

$$= .05(30)^2 + 2.4(30) - [.05(20)^2 + 2.4(20)]$$

$$= 49 \text{ trillion.}$$

43. This represents the increase in profits by increasing the production level from $x = 44$ to $x = 48$ units.

49. As shown in the last paragraph of p. 394, $A'(x) = f(x)$. Hence $A'(3) = f(3) = 3^2 + 1 = 10$.

6.4 Areas in the xy-Plane

1. $\int_{1}^{2} f(x)\, dx + \int_{3}^{4} [-f(x)]\, dx$ \hspace{2cm} (Fig. 10)

$\int_{2}^{3} f(x)\, dx + \int_{2}^{3} [-g(x)]\, dx$ \hspace{2cm} (Fig. 11)

7. A rough sketch of the curves are as follows,

$$y = x^2 - 6x + 12 = (x - 3)^2 + 3, \text{ which is } > 3. \quad \text{So,}$$

$$y = x^2 - 6x + 12 \text{ lies above } y = 1 \text{ for all } x.$$

Thus the area between the two curves is equal to

$$\int_{0}^{4} [(x^2 - 6x + 12) - (1)]\, dx$$

6-7

$$= \int_0^4 (x^2 - 6x + 11) \, dx = \frac{x^3}{3} - 3x^2 + 11x \Big|_0^4$$

$$= \frac{64}{3} - 48 + 44 = \frac{60}{3} = 20.$$

13. Only a rough sketch is needed. The graph of

$$y = -x^2 + 6x - 5$$

is obviously a parabola of some sort. The $-x^2$ term shows that the parabola opens down. (The second derivative of $-x^2 + 6x - 5$ is negative, which shows that the curve is concave down.) Also, the graph of $y = 3 - x$ is a line with negative slope. Without plotting any points or using a coordinate system, we can visualize the two possible locations of the parabola and line:

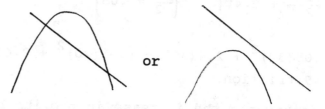

or

Hopefully, the line intersects the parabola. Otherwise there will be no region "bounded" by the graphs. To find the x-values of the intersection points, if any, we equate the y-values of the two curves and solve for x:

$$-x^2 + 6x - 5 = 2x - 5,$$
$$-x^2 + 4x = 0,$$
$$-x(x - 4) = 0,$$
$$x = 0, \text{ or } x = 4.$$

The rough sketch shows that the parabola is the top curve between $x = 0$ and $x = 4$. So the area bounded by the curves is

$$\int_0^4 [(-x^2 + 6x - 5) - (2x - 5)] \, dx$$

$$= \int_0^4 (-x^2 + 4x) \, dx = \frac{-x^3}{3} + 2x^2 \Big|_0^4 = \frac{-64}{3} + 32 = \frac{32}{3}.$$

19. The graph of $y = x^2 - 3x$ is obviously a parabola that opens upward. It crosses the x-axis at values of x that satisfy $x^2 - 3x = 0$, i.e., $x(x - 3) = 0$. Thus $x = 0$, or $x = 3$.

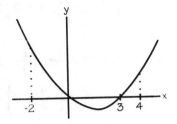

(a) The x-axis is the graph of the (constant) function $y = 0$. Since this graph is above the parabola from $x = 0$ to $x = 3$, the area of the region for part (a) is

$$\int_0^3 [0 - (x^2 - 3x)] \, dx = \int_0^3 3x - x^2 \, dx$$

$$= \frac{3}{2}x^2 - \frac{1}{3}x^3 \Big|_0^3 = (\frac{27}{2} - 9) - 0 = \frac{9}{2}.$$

(b) When x varies from $x = 0$ to $x = 4$, the two curves $y = 0$ and $y = 3x - x^2$ cross. We must consider two regions, one where $y = 0$ is on top and one where $y = 3x - x^2$ is on top. The first region was considered in part (a). The total area is equal to

$$[\text{Area found in (a)}] + \int_3^4 [(x^2 - 3x) - 0] \, dx$$

$$= \frac{9}{2} + \left[\frac{1}{3}x^3 - \frac{3}{2}x^2\right]\Big|_3^4$$

$$= \frac{9}{2} + (\frac{64}{3} - 24) - (9 - \frac{27}{2})$$

$$= \frac{27}{6} + \frac{128}{6} - 24 - 9 + \frac{81}{6}$$

$$= \frac{236}{6} - 33 = \frac{38}{6} = \frac{19}{3}.$$

(c) This is similar to part (b). When x is between -2 and 0, the graph of $y = 3x - x^2$ is on top; when x is between 0 and 3, the x-axis is on top. Thus the total area is

$$\int_{-2}^0 [(x^2 - 3x) - 0] \, dx + [\text{Area found in (a)}]$$

$$= \left(\frac{1}{3}x^3 - \frac{3}{2}x^2\right)\bigg|_{-1}^{0} + \frac{9}{2}$$

$$= 0 - \left(\frac{1}{3}(-2)^3 - \frac{3}{2}(-2)^2\right) + \frac{9}{2}$$

$$= -\left(-\frac{8}{3} - 6\right) + \frac{9}{2} = \frac{26}{3} + \frac{9}{2} = \frac{52}{6} + \frac{27}{6} = \frac{79}{6}.$$

25. $\begin{bmatrix} \text{amount of} \\ \text{depletion} \end{bmatrix} = \begin{bmatrix} \text{amount of} \\ \text{consumption} \end{bmatrix} - \begin{bmatrix} \text{amount of} \\ \text{new growth} \end{bmatrix}$

$$= \begin{bmatrix} \text{area of region} \\ \text{under consump-} \\ \text{tion curve} \end{bmatrix} - \begin{bmatrix} \text{area of region} \\ \text{under new} \\ \text{growth curve} \end{bmatrix}$$

$$= \begin{bmatrix} \text{area of region} \\ \text{between the curves} \end{bmatrix}$$

$$= \int_{0}^{20} [76.2e^{.03t} - (50 - 6.03e^{.09t})]\, dt$$

$$= \int_{0}^{20} (76.2e^{.03t} - 50 + 6.03e^{.09t})\, dt.$$

6.5 Applications of the Definite Integral

Although the derivations of the formulas presented in this section rely on Riemann sums, the exercises are worked by sub-stitution into the formulas and do not require analysis in terms of Riemann sums. Example 1 is important.

1. $f(x) = x^2$. By definition, the average value of $f(x)$ be-tween a and b is given by:

$$\text{Average value} = \frac{1}{b-a} \int_{a}^{b} f(x)\, dx$$

$$= \frac{1}{3-0} \int_{0}^{3} x^2\, dx = \frac{1}{3} \int_{0}^{3} x^2\, dx$$

$$= \frac{1}{3}\left[\frac{x^3}{3}\bigg|_{0}^{3}\right] = \frac{1}{3}\left[\frac{3^3}{3}\right] = 3.$$

7. $\text{Average temperature} = \frac{1}{12-0} \int_{0}^{12} \left(47 + 4t - \frac{1}{3}t^2\right) dt$

$$= \frac{1}{12}\left[47t + 2t^2 - \frac{1}{9}t^3 \Big|_0^{12}\right]$$

$$= \frac{1}{12}\left[47(12) + 2(12)^2 - \frac{1}{9}(12)^3\right]$$

$$= 47 + 2(12) - \frac{1}{9}(12)^2 = 55°.$$

13. Consumer's surplus $= \int_0^A [f(x) - B] \, dx, \quad B = f(A).$

$$f(x) = \frac{500}{x+10} - 3,$$

$$f(40) = \frac{500}{40+10} - 3 = \frac{500}{50} - 3 = 7,$$

so the surplus $= \int_0^{40} \left(\frac{500}{x+10} - 3 - 7\right) dx$

$$= 500 \int_0^{40} \frac{dx}{x+10} - \int_0^{40} 10 \, dx$$

$$= 500 \ln(x+10) \Big|_0^{40} - 10x \Big|_0^{40}$$

$$= 500 \ln 50 - 500 \ln 10 - 400$$

$$= 500 \ln 5 - 400$$

$$\approx \$404.72.$$

19. To find (A,B) set $12 - \frac{x}{50} = \frac{x}{20} + 5$

$$7 = \frac{x}{20} + \frac{x}{50} = \frac{7x}{100} \Rightarrow x = 100.$$

Thus A = 100, and B $= 12 - \frac{100}{50} = 10.$

To find the consumer's surplus $= \int_0^{100} \left(12 - \frac{x}{50} - 10\right) dx$

$$= \int_0^{100} \left(2 - \frac{x}{50}\right) dx = 2x - \frac{x^2}{100} \Big|_0^{100}$$

$$= 200 - 100 = \$100.$$

Producers surplus $= \int_0^A (f(A) - f(x)) \, dx$

$$= \int_0^{100} \left[10 - \left(\frac{x}{20} + 5 \right) \right] dx$$

$$= \int_0^{100} \left(5 - \frac{x}{20} \right) dx = 5x - \frac{x^2}{40} \Big|_0^{100}$$

$$= 500 - \frac{10000}{40} = \$250.$$

25. Future amount $= \int_0^N P\, e^{r(N-t)}\, dt.$

We don't know N, and this is what we wish to evaluate. So

$$140,000 = \int_0^N 5000\, e^{.1(N-t)}\, dt$$

$$= 5000 \int_0^N e^{.1N}\, e^{-.1t}\, dt$$

hence,

$$\frac{240,000}{5,000} = 28 = \int_0^N e^{.1N}\, e^{-.1t}\, dt$$

$$= e^{.1N} \int_0^N e^{-.1t}\, dt = e^{.1N} \left[\frac{e^{-.1t}}{-.1} \Big|_0^N \right]$$

$$= -10e^{.1N} \left[e^{-.1N} - 1 \right].$$

So,

$$28 = -10 + 10e^{.1N},$$

$$38 = 10e^{.1N},$$

$$3.8 = e^{.1N}.$$

Taking the logarithm of both sides, we have $\ln 3.8 = .1N$. Then $10 \ln 3.8 = N$ and $N \approx 13.35$ years.

31. The volume is given by

$$\int_1^2 \pi (x^2)^2\, dx = \int_1^2 \pi x^4\, dx = \frac{\pi}{5} x^5 \Big|_1^2 = \frac{31\pi}{5} \text{ cubic units.}$$

37. The Riemann sum tells us that $f(x) = x^3$, $n = 4$, and $\Delta x = .5$. Since $a = 8$ and $\Delta x = \frac{b-a}{n}$, we have

$$.5 = \frac{b-8}{4},$$

so

$$2 = b-8, \text{ or } b = 10.$$

(for *Calculus and Its Applications*, 7th Edition.)

1. Since $\frac{d}{dx}(-2e^{-x/2}) = e^{-x/2}$, we have

$$\int e^{-x/2} \, dx = -2e^{-x/2} + C.$$

7. $\int_1^4 \frac{1}{x^2} \, dx = \int_1^4 x^{-2} \, dx = -x^{-1} \Big|_1^4 = -\frac{1}{4} + 1 = \frac{3}{4}.$

Warning: A common error is to think that $\frac{1}{x^3}$ is an antiderivative of $\frac{1}{x^2}$. You avoid this mistake when you switch to the
negative exponent notation, $\frac{1}{x^2} = x^{-2}$.

13. A sketch of the graph reveals three points of intersection. Equating the two expressions in x we obtain

$$x^3 - 3x + 1 = x + 1,$$
$$x^3 - 4x = 0,$$
$$x(x^2 - 4) = 0,$$
$$x(x-2)(x+2) = 0.$$

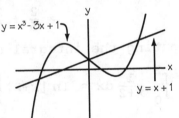

Hence, x = 0, 2, or -2. The total area bounded by the
curves is the sum of the areas of the two regions.

$$\text{Area} = \int_{-2}^0 [(x^3-3x+1) - (x+1)] \, dx + \int_0^2 [(x+1) - (x^3-3x+1)] \, dx$$

$$= \int_{-2}^0 (x^3 - 4x) \, dx + \int_0^2 (-x^3 + 4x) \, dx$$

$$= (\tfrac{1}{4}x^4 - 2x^2) \Big|_{-2}^0 + (-\tfrac{1}{4}x^4 + 2x^2) \Big|_0^2$$

$$= 0 - (4 - 8) + (-4 + 8) - 0$$

$$= 8.$$

19. The cost of producing x tires a day is an antiderivative
of the marginal cost function .04x + 150. Hence, if we
denote the cost function by C(x), we have

$$C(x) = \int (.04x + 150)\, dx = .02x^2 + 150x + C.$$

If fixed costs are $500 per day, then the cost of producing no tires is $500 per day. That is,

$$C(0) = 0 + 0 + C = C = 500.$$

Hence

$$C(x) = .02x^2 + 150x + 500.$$

25. We have $\Delta x = \dfrac{b-a}{n} = \dfrac{2-0}{2} = 1$, so the midpoints of the two subintervals are given by

$$x_1 = a + \frac{\Delta x}{2} = 0 + \frac{1}{2} = \frac{1}{2},$$

$$x_2 = \frac{1}{2} + \Delta x = \frac{1}{2} + 1 = \frac{3}{2}.$$

Using the midpoint rule, we have

$$\text{Area} \approx [f(\tfrac{1}{2}) + f(\tfrac{3}{2})](1) = \left[\frac{1}{1/2 + 2} + \frac{1}{3/2 + 2} \right]$$

$$= \frac{2}{5} + \frac{2}{7} = \frac{24}{35} \approx .68571.$$

Computing the integral directly, we have

$$\int_0^2 \frac{1}{x+2}\, dx \approx \ln |x+2| \Big|_0^2$$

$$= \ln 4 - \ln 2 = \ln \frac{4}{2} = \ln 2 = .69315.$$

31. Observe that the sum is a Riemann sum (using left endpoints of the subintervals) for the function $f(x) = e^x - \dfrac{3}{x}$ from $x = 2$ to $x = 3$ with $\Delta x = .01$. This sum is approximately equal to the value of the definite integral

$$\int_2^3 (e^x - \frac{3}{x})\, dx = (e^x - 3\ln|x|) \Big|_2^3$$

$$= [e^3 - 3\ln 3] - [e^2 - 3\ln 2]$$

$$\approx 20.086 - 3.296 - 7.389 + 2.079$$

$$= 11.48.$$

37. Observe that the sum is a Riemann sum for the function $f(t) = 5000e^{-.1t}$ from $t = 0$ to $t = 3$. This sum is approximately equal to the value of the definite integral:

$$\int_0^3 5000e^{-.1t}\,dt = -50{,}000e^{-.1t}\Big|_0^3$$

$$= -50{,}000e^{-.3} + 50{,}000$$

$$\approx 13{,}000.$$

43. Amount of water used between 1940 and 1980:

$$= \int_0^{40} 860\,e^{.04t}\,dt$$

$$= 860\,\frac{e^{.04t}}{.04}\bigg|_0^{40} = 21{,}500\left(e^{.04(40)} - e^0\right)$$

$$= 21{,}500\left(e^{1.6} - 1\right) \approx 84{,}990.2 \text{ km}^3.$$

6.6 Techniques of Integration (Brief Edition)

The technique of integration by substitution is best learned by trial and error. The more problems you work the fewer errors you will make in your first guess at a substitution. The basic strategy is to arrange the integrand in the form $f(g(x))g'(x)$. Look for the most complicated part of the integrand that may be expressed as a composite function, and let u be the "inside" function $g(x)$.

Exercises 21-30 are all solved using integration by parts. Once you have mastered integration by parts and by substitution, you are ready to try Exercises 31-46. These problems are harder because you have to decide which method to use!

a. Always consider the method of substitution first. This will work if the integrand is the product of two functions and one of the functions is the derivative of the "inside" of the other function (except maybe for a constant multiple).

b. If the integrand is the product of two unrelated functions, try integration by parts.

c. Some antiderivatives cannot be found by either of the methods we have discussed. We will never give you such a problem to solve, but you should be careful if you look in another text for more problems to work.

Remember that you can check your answers to an antidifferentiation problem by differentiating your answer. (You ought to do this at least mentally for every indefinite integral you find.) We urge you to work *all* the problems (odd and even) in Section 6.6. Use the solutions in this guide to check your work, *not* to show you how to start a problem.

1. Let $u = x^2 + 4$. Then $du = \frac{d}{dx}(x^2+4)\,dx = 2x\,dx$.

$$\int 2x(x^2+4)^5\,dx = \int u^5\,du$$

$$= \frac{1}{6}u^6 + C$$

$$= \frac{1}{6}(x^2+4)^6 + C.$$

7. Let $u = x^3 - 1$. Then $du = 3x^2\,dx$, and

$$\int \frac{3x^2}{x^3-1}\,dx = \int \frac{1}{x^3-1}\cdot 3x^2\,dx = \int \frac{1}{u}\,du$$

$$= \ln|u| + C = \ln|x^3-1| + C.$$

13. Let $u = x^3 + 3x + 2$. Then $du = 3x^3 + 3\,dx$, and

$$\int \frac{x^2+1}{x^3+3x+2}\,dx = \frac{1}{3}\int \frac{1}{x^3+3x+2}\cdot 3(x^2+1)\,dx = \frac{1}{3}\int \frac{1}{u}\,du$$

$$= \frac{1}{3}\ln|u| + C = \frac{1}{3}\ln|x^3+3x+2| + C.$$

19. Let $u = -x^2$. Then $du = -2x\,dx$, and

$$\int \frac{x}{e^{x^2}}\,dx = \int xe^{-x^2}\,dx = \frac{1}{-2}\int e^{-x^2}\cdot(-2x)\,dx$$

$$= -\frac{1}{2}\int e^u\,du = -\frac{1}{2}e^u + C = -\frac{1}{2}e^{-x^2} + C.$$

25. Let $f(x) = x$, $\quad g(x) = \sqrt{x+1} = (x+1)^{1/2}$,

$\quad\quad f'(x) = 1$, $\quad G(x) = \frac{2}{3}(x+1)^{3/2}$.

Then $\int x\sqrt{x+1}\,dx = \frac{2}{3}x(x+1)^{3/2} - \int \frac{2}{3}(x+1)^{3/2}\,dx$

$$= \frac{2}{3}x(x+1)^{3/2} - \frac{2}{3}\int (x+1)^{3/2}\,dx$$

$$= \frac{2}{3}x(x+1)^{3/2} - \frac{2}{3}\cdot\frac{2}{5}(x+1)^{5/2} + C$$

$$= \frac{2}{3}x(x+1)^{3/2} - \frac{4}{15}(x+1)^{5/2} + C.$$

31. Since xe^{2x} is the product of two unrelated functions, we try integration by parts. Let

$$f(x) = x, \quad g(x) = e^{2x},$$

$$f'(x) = 1, \quad G(x) = \tfrac{1}{2}e^{2x}.$$

Then $\displaystyle \int xe^{2x}\,dx = \tfrac{1}{2}xe^{2x} - \int \tfrac{1}{2}e^{2x}\,dx$

$$= \tfrac{1}{2}xe^{2x} - \tfrac{1}{2}\int e^{2x}\,dx$$

$$= \tfrac{1}{2}xe^{2x} - \tfrac{1}{2}\cdot\tfrac{1}{2}e^{2x} + C$$

$$= \tfrac{1}{2}xe^{2x} - \tfrac{1}{4}e^{2x} + C.$$

Helpful Hint: The various steps in your calculation of an integral such as $\int xe^{2x}\,dx$ should be connected by equals signs, showing how you move from one step to the next. It requires a little more effort to rewrite the first term $\tfrac{1}{2}xe^{2x}$ on each line, but the solution is not correct otherwise. A careless student might write

$$\int xe^{2x}\,dx = \tfrac{1}{2}xe^{2x} - \int \tfrac{1}{2}e^{2x}\,dx$$

$$\boxed{=} \qquad\qquad - \tfrac{1}{2}\int e^{2x}\,dx$$

$$\boxed{\text{errors}} \rightarrow \boxed{=} \qquad\qquad - \tfrac{1}{2}\cdot\tfrac{1}{2}e^{2x} + C$$

$$\boxed{=} \qquad\qquad - \tfrac{1}{4}e^{2x} + C.$$

Avoid such "shortcuts" in your *homework* as well as on tests. Develop the habit of writing proper mathematical solutions on your homework. This will help to generate correct patterns of thought when you take an exam.

37. Let $u = 6x^2 + 9x$, then $du = (12x + 9)\,dx = 3(4x+3)\,dx$.

$$\int (4x+3)(6x^2+9x)^{-7}\,dx = \tfrac{1}{3}\int (6x^2+9x)^{-7}\cdot 3(4x+3)\,dx$$

$$= \tfrac{1}{3}\int u^{-7}\,du = \tfrac{1}{3}\left(-\tfrac{1}{6}u^{-6}\right) + C$$

$$= -\tfrac{1}{18}(6x^2+9x)^{-6} + C.$$

43. Use the formula for the present value P of an income stream with K(t) = 300t - 500, T = 5 years, and r = .10.

$$P = \int_0^5 (300t-500) e^{-.1t} dt.$$

We use integration by parts, with

$$f(t) = 300t - 500, \quad g(t) = e^{-.1t},$$
$$f'(t) = 300, \quad G(t) = -10e^{-.1t}.$$

Then

$$\int (300t-500) e^{-.1t} dt = (300t-500)(-10e^{-.1t})$$
$$- \int 300(-10e^{-.1t}) dt$$
$$= (5000-3000t) e^{-.1t} + 3000 \int e^{-.1t} dt$$
$$= (5000-3000t) e^{-.1t} + 3000(-10e^{-.1t})$$
$$= (5000-3000t-30,000) e^{-.1t}$$
$$= (-25,000-3000t) e^{-.1t}.$$

Then

$$P = (-25,000-3000t) e^{-.1t} \Big|_0^5$$
$$= (-25,000-15,000) e^{-.5} - (-25,000) e^0$$
$$= -40,000(.60653) + 25,000$$
$$= -24,261.20 + 25,000 = \$738.80.$$

6.7   Improper Integrals (Brief Edition)

This section in the Brief Edition is the same as Section 9.6 in *Calculus and Its Applications, 7th edition*. The remarks and exercise solutions for this section may be found later in the *Study Guide*, with the other material for Chapter 9.

6.8   Applications of Calculus to Probability (Brief Edition)

A probability density function on an interval A ≤ x ≤ B is a function f(x) such that, for A ≤ x ≤ B, the graph of f(x) lies above or on the x-axis and the area of the region under the graph equals 1.

1. Let X be the outcome of an experiment whose probability density function is $f(x) = 6(x-x^2) = 6x - 6x^2$.

a. $Pr(\frac{1}{4} \le X \le \frac{1}{2}) = \int_{1/4}^{1/2} (6x-6x^2)\, dx = (3x^2-2x^3)\Big|_{1/4}^{1/2}$

$$= [3(\tfrac{1}{2})^2 - 2(\tfrac{1}{2})^3] - [3(\tfrac{1}{4})^2 - 2(\tfrac{1}{4})^3]$$

$$= \frac{1}{2} - \frac{10}{64} = \frac{22}{64} = \frac{11}{32}.$$

b. $Pr(0 \le X \le \frac{1}{3}) = \int_0^{1/3} (6x-6x^2)\, dx = (3x^2-2x^3)\Big|_0^{1/3}$

$$= [3(\tfrac{1}{3})^2 - 2(\tfrac{1}{3})^3] - 0$$

$$= \frac{3}{9} - \frac{2}{27} = \frac{7}{27}.$$

c. Since the values of X lie between 0 and 1, the probability that X is at least 1/4 is the same (by definition) as the probability that X is between 1/4 and 1.

$$Pr(\tfrac{1}{4} \le X) = Pr(\tfrac{1}{4} \le X \le 1) = \int_{1/4}^{1} (6x-6x^2)\, dx$$

$$= (3x^2 - 2x^3)\Big|_{1/4}^{1} = [3-2] - [3(\tfrac{1}{4})^2 - 2(\tfrac{1}{4})^3]$$

$$= 1 - [\frac{3}{16} - \frac{2}{64}] = 1 - \frac{10}{64} = \frac{54}{64} = \frac{27}{32}.$$

d. $Pr(X \le \frac{3}{4}) = Pr(0 \le X \le \frac{3}{4}) = \int_0^{3/4} (6x-6x^2)\, dx.$

$$= (3x^2-2x^3)\Big|_0^{3/4} = [3(\tfrac{3}{4})^2 - 2(\tfrac{3}{4})^3] - 0$$

$$= \frac{27}{16} - \frac{27}{32} = \frac{27}{32}.$$

7. **a.** The fact that the value of X is determined by a "random" arrival time means that X has a uniform probability density function, $f(x) = k$, where k is a suitable constant. The values of X lie between 0 and 3 minutes. We must choose k so that the area of the region under the graph of $f(x)$ is 1. Since the graph is a horizontal line at height k, over the interval $0 \le x \le 3$, the region is a rectangle with area [base]·[height] = 3·k. So we need $3k = 1$ and $k = 1/3$. Thus

$$f(x) = \frac{1}{3}, \quad 0 \le x \le 3.$$

b. Since the values of X do not exceed 3,

$$\Pr(1 \leq X) = \Pr(1 \leq X \leq 3) = \int_{1}^{3} \frac{1}{3}\,dx = \frac{1}{3}x\Big|_{1}^{3}$$

$$= \frac{1}{3}(3) - \frac{1}{3}(1) = 1 - \frac{1}{3} = \frac{2}{3}.$$

c. Since the values of x are not less than 0,

$$\Pr(X \leq 1) = \Pr(0 \leq X \leq 1) = \int_{0}^{1} \frac{1}{3}\,dx = \frac{1}{3}x\Big|_{0}^{1}$$

$$= \frac{1}{3}(1) - 0 = \frac{1}{3}.$$

13. The exponential probability density function for x has the form $f(x) = \lambda e^{-\lambda x}$. We find λ from the fact that

$$\lambda = \frac{1}{[\text{average value of X}]}.$$

In this exercise, the average service time is 4 minutes, so $\lambda = 1/4$ and

$$f(x) = \frac{1}{4}e^{-(1/4)x} = \frac{1}{4}e^{-x/4}.$$

a. $\Pr(X \leq 2) = \Pr(0 \leq X \leq 2) = \int_{0}^{2} \frac{1}{4}e^{-x/4}\,dx$

$$= -e^{-x/4}\Big|_{0}^{2} = -e^{-2/4} - (-e^{0})$$

$$= 1 - e^{-1/2} = 1 - .60657 = .39347.$$

b. $\Pr(4 \leq X) = \Pr(4 \leq X < \infty) = \int_{4}^{\infty} \frac{1}{4}e^{-x/4}\,dx.$

We have

$$\int_{4}^{b} \frac{1}{4}e^{-x/4}\,dx = -e^{-x/4}\Big|_{4}^{b} = -e^{-b/4} - (-e^{-4/4})$$

$$= e^{-1} - e^{-b/4} \to e^{-1} \text{ as } b \to \infty.$$

Thus,

$$\Pr(4 \leq X) = \int_{4}^{\infty} \frac{1}{4}e^{-x/4}\,dx = e^{-1} = .36788.$$

Chapter 6: Supplementary Exercises (Brief Edition)

1. By inspection, $\int e^{-x/2}\,dx = -2e^{-x/2} + C$. Always remember to check your answer.

7. The second function in the integrand is the derivative of the "inside" of the first function, so we use the substitution $u = e^x + 4$, $du = e^x\,dx$:

$$\int (e^x + 4)^3 \cdot e^x\,dx = \int u^3\,du = \tfrac{1}{4}u^4 + C = \tfrac{1}{4}(e^x + 4)^4 + C.$$

13. Since $x(2x + 3)^2$ is the product of two unrelated functions, we use integration by parts. Let

$$f(x) = x, \quad g(x) = (2x + 3)^7,$$
$$f'(x) = 1, \quad G(x) = \tfrac{1}{8 \cdot 2}(2x + 3)^8 = \tfrac{1}{16}(2x + 3)^8.$$

Don't forget to check that $G(x)$ is a correct antiderivative of $g(x)$. (Differentiate $G(x)$.) Then

$$\int x(2x + 3)^7\,dx = x \cdot \tfrac{1}{16}(2x + 3)^8 - \int \tfrac{1}{16}(2x + 3)^8\,dx$$

$$= \tfrac{x}{16}(2x + 3)^8 - \tfrac{1}{16} \cdot \tfrac{1}{9 \cdot 2}(2x + 3)^9 + C$$

$$= \tfrac{x}{16}(2x + 3)^8 - \tfrac{1}{288}(2x + 3)^9 + C.$$

19. Since x^4 is almost (up to a multiple) the derivation of $x^5 - 1$, we use the substitution $u = x^5 - 1$, $du = 5x^4$. Then

$$\int x^4 e^{x^5 - 1}\,dx = \tfrac{1}{5}\int 5x^4 e^{x^5 - 1}\,dx = \tfrac{1}{5}\int e^u\,du$$

$$= \tfrac{1}{5}e^u + C = \tfrac{1}{5}e^{x^5 - 1} + C.$$

25. The graph of $y = 16 - x^2$ is a parabola that opens downward. The graph of $y = 10 - t$ is a line with negative slope. Since the two curves determine a bounded region, the two curves must intersect. The general shape of the region is shown at the right. Clearly the parabola lies above the line for this region. We find the x-coordinates of the intersec-

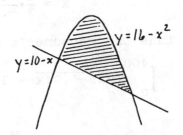

tion points by solving the equation:
$$10 - x = 16 - x^2,$$
$$x^2 - x - 6 = 0,$$
$$(x-3)(x+2) = 0.$$

Hence $x = -2$ or $x = 3$. The area of the region is

$$\int_{-2}^{3} [(16 - x^2) - (10 - x)]\, dx = \int_{-2}^{3} (-x^2 + x + 6)\, dx$$

$$= (-\tfrac{1}{3}x^3 + \tfrac{1}{2}x^2 + 6x)\Big|_{-2}^{3}$$

$$= [-9 + \tfrac{9}{2} + 18] - [\tfrac{8}{3} + 2 - 12]$$

$$= \frac{125}{6}.$$

31. With the demand curve $p = \sqrt{25 - .04x}$, the sales level $x = 400$ corresponds to the price $p = \sqrt{25 - .04(400)} = \sqrt{25 - 16} = 3$. The consumer's surplus is the area of the region below the demand curve and above the horizontal line $p = 3$, from $x = 0$ to $x = 400$. This area is given by

$$\int_{0}^{400} (\sqrt{25 - .04x} - 3)\, dx.$$

You should be able by now to guess that an antiderivative of $(25 - .04x)^{1/2}$ has the form $k(25 - .04x)^{3/2}$. To find k, we differentiate

$$\frac{d}{dx} k(25 - .04x)^{3/2} = \frac{3}{2} k(25 - .04x)^{1/2} \cdot (-.04).$$

In order for this derivative to be exactly $(25 - .04x)^{1/2}$, we need

$$\frac{3}{2} k(-.04) = 1,$$

$$k = \frac{2}{3} \cdot \frac{1}{-.04} = -\frac{2}{3} \cdot 25 = -\frac{50}{3}.$$

Thus the consumer's surplus equals

$$\int_{0}^{400} (\sqrt{25 - .04x} - 3)\, dx = \left[-\frac{50}{3}(25 - .04x)^{3/2}) - 3x \right]\Big|_{0}^{400}$$

$$= \left[-\frac{50}{3}(9)^{3/2} - 1200 \right] - \left[-\frac{50}{3}(25)^{3/2} - 0 \right]$$

$$= \left[-\frac{50}{3}(27) - 1200\right] + \frac{50}{3}(125)$$

$$= -1650 + \frac{6250}{3} = \frac{1300}{3} = 433.33.$$

The consumer's surplus is \$433.33.

37. Suppose that $y = f(t)$ has the property that $y' = kty$, that is,

$$f'(t) = ktf(t), \quad \text{for all } t. \qquad (*)$$

Following the hint in the problem, we use the product rule to compute the derivative:

$$\frac{d}{dt}\left[f(t)e^{-kt^2/2}\right] = f(t) \cdot \frac{d}{dt}e^{-kt^2/2} + e^{-kt^2/2} \cdot f'(t)$$

$$= f(t)e^{-kt^2/2} \cdot (-kt) + e^{-kt^2/2} \cdot f'(t)$$

$$= e^{-kt^2/2}[-ktf(t) + f'(t)].$$

The term on the right in brackets must be zero for all t, because $f(t)$ satisfies (*) above. So

$$\frac{d}{dt}\left[f(t)e^{-kt^2/2}\right] = 0, \quad \text{for all } t.$$

By Theorem II of Section 6.1, a function whose derivative is zero for all t must itself be a constant function. So there is a constant C such that

$$f(t)e^{-kt^2/2} = C, \quad \text{for all } t.$$

Solving for $f(t)$, we conclude that $f(t) = Ce^{-kt^2/2}$.

43. For any $b > 0$, we compute

$$\int_0^b e^{6-3x}\,dx = -\frac{1}{3}e^{6-3x}\Big|_0^b$$

$$= [-\frac{1}{3}e^{6-3b}] - [-\frac{1}{3}e^6] = \frac{1}{3}e^6 - \frac{1}{3}e^{6-3b}.$$

We observe that

$$\frac{1}{3}e^{6-3b} = \frac{e^6}{3e^{3b}} \to 0 \text{ as } b \to \infty.$$

Thus

$$\int_0^\infty e^{6-3x}\,dx = \lim_{b\to\infty} \int_0^b e^{6-3x}\,dx = \lim_{b\to\infty} \left[\frac{1}{3}e^6 - \frac{1}{3}e^{6-3b}\right]$$

$$= \frac{1}{3}e^6.$$

49. The average life span (or mean) is given by $E(X) = \frac{1}{k} = 72$. Hence $k = \frac{1}{72}$, and the probability density function is $f(x) = \frac{1}{72}e^{-(1/72)x}$. The probability that a component lasts for more than 24 months is given by

$$Pr(24 \leq X) = \int_{24}^\infty \frac{1}{72}e^{-(1/72)x}\,dx$$

$$= \lim_{b\to\infty} \int_{24}^b \frac{1}{72}e^{-(1/72)x}\,dx$$

$$= \lim_{b\to\infty} \left[\left(-e^{-(1/72)x}\right)\Big|_{24}^b\right]$$

$$= \lim_{b\to\infty} \left[e^{-1/3} - e^{-b/72}\right]$$

$$= e^{-1/3}.$$

CHAPTER 7

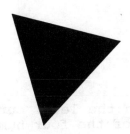

FUNCTIONS OF SEVERAL VARIABLES

7.1 Examples of Functions of Several Variables

You should be sure that you completely understand the meanings of the variables in Examples 2 and 3. These examples will be used to illustrate the concepts introduced in later sections.

1. $f(x,y) = x^2 + 8y$. So

$$f(1,0) = (1)^2 + (8)(0) = 1,$$
$$f(0,1) = (0)^2 + (8)(1) = 8,$$
$$f(3,2) = (3)^2 + (8)(2) = 9 + 16 = 25.$$

7. $f(x,y) = xy$. So

$$f(2+h,3) - f(2,3) = (2+h)3 - (2)(3)$$
$$= 6 + 3h - 6$$
$$= 3h.$$

13. (a) $v = 200,000,$
    $x = 5000,$
    $r = 2.5,$

$$T = \frac{r}{100}[.4v - x],$$

$$T = \frac{2.5}{100}[.4(200,000) - 5000]$$

$$= \$1875.00.$$

(b) If $r = 3.00,$

$$T = \frac{3.00}{100}[.4(200,000) - 5000]$$

$$= \$2250.$$

The tax due also increases by 20% (or $\frac{1}{5}$), since

$$1875.00 + \frac{1}{5}(1875.00) = \$2250.00$$

19. A level curve has the form $f(x,y) = C$ (C is arbitrary). Rearrange the equation $y = 3x - 4$ so that all terms of x and y are on the left-hand side:

$$y - 3x = -4.$$

So,

$$y - 3x = C, \text{ some constant.}$$

Thus $f(x,y) = y - 3x$.

25. The graph in Exercise 25 is matched with the level curves in (c). Imagine slicing "near the top" of the four humps. We obtain a cross section of four circular-like figures and as we move further down the z-axis these circular-like figures become larger and larger.

Similarly, Exercise 23 is matched with (d),
        Exercise 24 is matched with (b),
        Exercise 26 is matched with (a).

7.2 Partial Derivatives

Any letter can be used as a variable of differentiation, even y. As a warm-up to this section, look over the following ordinary derivatives.

$$\text{If } f(y) = ky^2, \text{ then } \frac{df}{dy} = 2ky.$$

$$\text{If } g(y) = 3y^2x, \text{ then } \frac{dg}{dy} = 6yx.$$

Notice that x is a constant in the formula for $g(y)$ because the notation $g(y)$ means that y is the only variable in the function.

1. $\frac{\partial}{\partial x} 5xy = \frac{\partial}{\partial x} [5y]x = 5y$, [we treat $5y$ as a constant]

$\frac{\partial}{\partial y} 5xy = \frac{\partial}{\partial y} [5x]y = 5x.$ [we treat $5x$ as a constant]

7. $\frac{\partial}{\partial x}(2x-y+5)^2 = 2(2x-y+5) \cdot \frac{\partial}{\partial x}(2x-y+5)$

$$= 2(2x-y+5) \cdot (2-0+0) \quad \text{[we treat } y \text{ as a constant]}$$

$$= 4(2x-y+5).$$

$\frac{\partial}{\partial y}(2x-y+5)^2 = 2(2x-y+5) \cdot \frac{\partial}{\partial y}(2x-y+5)$

$$= 2(2x-y+5) \cdot (0-1+0) \quad \text{[we treat } 2x \text{ as a constant]}$$

$$= -2(2x-y+5).$$

7-2

13. $f(L,K) = 3\sqrt{LK} = 3(LK)^{1/2} = 3L^{1/2}K^{1/2}$. So

$$\frac{\partial}{\partial L} 3L^{1/2}K^{1/2} = \frac{\partial}{\partial L}[3K^{1/2}]L^{1/2} = [3K^{1/2}]\frac{1}{2}L^{-1/2}$$

$$= \frac{3}{2}K^{1/2}L^{-1/2}$$

$$= \frac{3\sqrt{K}}{2\sqrt{L}}.$$

19. $\frac{\partial}{\partial x}(x^2 + 2xy + y^2 + 3x + 5y) = 2x + 2y + 0 + 3 + 0$

$$= 2x + 2y + 3.$$

Thus $\frac{\partial f}{\partial x}(2,-3) = 2(2) + 2(-3) + 3 = 4 - 6 + 3 = 1$.

$\frac{\partial}{\partial y}(x^2 + 2xy + y^2 + 3x + 5y) = 0 + 2x + 2y + 0 + 5$

$$= 2x + 2y + 5.$$

Thus $\frac{\partial f}{\partial y}(2,-3) = 2(2) + 2(-3) + 5 = 4 - 6 + 5 = 3$.

25. (a) Marginal productivity of labor is given by $\frac{\partial f}{\partial x}$.

$$f(x,y) = 200\sqrt{6x^2+y^2} = 200(6x^2 + y^2)^{1/2}.$$

$$\frac{\partial f}{\partial x} = 200(\frac{1}{2})(6x^2+y^2)^{-1/2} \cdot \frac{\partial}{\partial x}(6x^2+y^2)$$

$$= 100(6x^2+y^2)^{-1/2}(12x) = \frac{1200x}{\sqrt{6x^2+y^2}}.$$

$$\frac{\partial f}{\partial x}(10,5) = \frac{1200(10)}{\sqrt{6(10)^2+(5)^2}} = \frac{12,000}{\sqrt{625}} = \frac{12,000}{25} = 480.$$

Marginal productivity of capital is given by $\frac{\partial f}{\partial y}$.

$$\frac{\partial f}{\partial y} = 200(\frac{1}{2})(6x^2+y^2)^{-1/2} \cdot \frac{\partial}{\partial y}(6x^2+y^2)$$

$$= 100(6x^2+y^2)^{-1/2} \cdot (2y) = \frac{200y}{\sqrt{6x^2+y^2}}.$$

$$\frac{\partial f}{\partial y}(10,5) = \frac{200(5)}{\sqrt{6(10)^2+(5)^2}} = \frac{1000}{\sqrt{625}} = \frac{1000}{25} = 40.$$

(b) If capital is fixed at 5 units and labor is decreased by one unit from 10 to 9 then the quantity of goods produced will decrease by approximately 480 units of goods. Thus if capital is fixed at 5 units, and labor decreased by 1/2 unit from 10 to 9.5 units, the number of goods produced will decrease by approximately 480/2 or 240 units of goods.

31. $V = \dfrac{.08T}{P} = .08TP^{-1}$,

$\dfrac{\partial V}{\partial P} = -(.08)TP^{-2}$,

$\dfrac{\partial V}{\partial P}(20,300) = -(.08)(300)(20)^{-2} = \dfrac{-(.08)(300)}{400} = -.06$,

$\dfrac{\partial V}{\partial T} = .08P^{-1}$,

$\dfrac{\partial V}{\partial T}(20,300) = .08(20)^{-1} = \dfrac{.08}{20} = .004$,

$\dfrac{\partial V}{\partial P}(20,300)$ represents the rate of change of volume, when temperature is fixed at 300 and pressure is allowed to vary near 20. That is, if the pressure is increased by one small unit, volume will decrease by about .06 units.

$\dfrac{\partial V}{\partial T}(20,300)$ represents the rate of change of volume when pressure is fixed at 20 and temperature is allowed to vary near 300. That is, if temperature is increased by one small unit, volume will increase by approximately .004 units.

37. $f(x,y) = 3x^2 + 2xy + 5y$,

$f(1+h,4) - f(1,4) = 3(1+h)^2 + 2(1+h)(4) + 5(4)$

$\qquad\qquad\qquad\qquad -[3(1)^2 + 2(1)(4) + 5(4)]$

$\qquad = 3(1+h^2+2h) + 8 + 8h + 20 - 3 - 8 - 20$

$\qquad = 3 + 3h^2 + 6h + 8h - 3$

$\qquad = 3h^2 + 14h$.

7.3 Maxima and Minima of Functions of Several Variables

Finding relative extreme points for functions of several variables is accomplished in much the same way as for functions of one variable. In Chapter 2 we found relative maximum and minimum points by differentiating and then solving one equation for x. In this section, we differentiate (with partials) and then solve a system of two equations in two unknowns.

Three types of systems of equations arise in this section.

First type: Each equation involves just one variable.

An example is:

$$\begin{cases} x^2 + 1 = 5, \\ 3y - 4 = 11. \end{cases}$$

To solve this system, solve each equation separately for its variable. The solutions to the system consist of all combinations of these individual solutions. In the example above, the first equation has the two solutions $x = 2$ and $x = -2$. The second equation has the single solution $y = 5$. Therefore, the system has two solutions: $x = 2$, $y = 5$ and $x = -2$, $y = 5$. [Note: These two solutions can also be written $(2,5)$ and $(-2,5)$.]

Second type: One equation contains both variables, the other equation contains just one variable.

An example is

$$\begin{cases} 2x + 4y = 26, \\ 3y - 4 = 11. \end{cases}$$

To solve this system, solve the equation having just one variable, substitute this value into the other equation, and then solve that equation for the remaining variable. In the example above, the second equation has the solution $y = 5$. Substituting into the first equation, we obtain $2x + 4(5) = 26$ or $x = 3$. Therefore, the system has the solution $x = 3$, $y = 5$.

Third type: Both equations contain both variables.

An example is

$$\begin{cases} -9x + y = 3, \\ y + 2 = 4x. \end{cases}$$

Two methods for solving such systems are the EQUATE METHOD and the SOLVE-SUBSTITUTE METHOD.

With the EQUATE METHOD, both equations are solved for y, the two expressions for y are equated and solved for x, and the value of x is substituted into one of the equations to obtain the value of y. The solution to the system above is

$$\begin{cases} y = 3 + 9x, \\ y = 4x - 2. \end{cases}$$

Equating

$$3 + 9x = 4x - 2,$$
$$5x = -5,$$
$$x = -1.$$

Substituting into the first equation, $y = 3 + 9(-1) = -6$. Therefore, the solution is $x = -1$, $y = -6$.

With the SOLVE-SUBSTITUTE METHOD, one equation is solved for y as an expression in x, this expression is substituted into the other equation, that equation is solved for x, and the value(s) of x is substituted into the expression for y to obtain the value(s) of y. The solution to the system above is found as follows:

Solve first equation for y: $y = 3 + 9x$.
Substitute into 2nd eqn: $(3 + 9x) + 2 = 4x$.
Solve for x: $x = -1$.

Substitute the x value into the equation for y:

$$y = 3 + 9(-1) = -6.$$

1. $\frac{\partial}{\partial x}(x^2 - 3y^2 + 4x + 6y + 8) = 2x - 0 + 4 + 0 + 0 = 2x + 4,$

$\frac{\partial}{\partial y}(x^2 - 3y^2 + 4x + 6y + 8) = 0 - 6y + 0 + 6 + 0 = -6y + 6.$

To find possible relative maxima or minima, set the first partial derivatives equal to zero.

$\frac{\partial f}{\partial x} = 2x + 4 = 0,$ $\qquad\qquad$ $\frac{\partial f}{\partial y} = -6y + 6 = 0,$

$\qquad 2x = -4,$ $\qquad\qquad\qquad\qquad\qquad -6y = -6,$

$\qquad x = -4/2 = -2,$ $\qquad\qquad\qquad\qquad y = 1.$

Thus $f(x,y)$ has a possible relative maximum or minimum at $(-2,1)$.

7. $\frac{\partial}{\partial x}(\frac{1}{3}x^3 - 2y^3 - 5x + 6y - 5) = x^2 - 0 - 5 + 0 - 0 = x^2 - 5,$

$\frac{\partial}{\partial y}(\frac{1}{3}x^3 - 2y^3 - 5x + 6y - 5) = 0 - 6y^2 - 0 + 6 - 0$

$$= -6y^2 + 6.$$

Again, we set the first partial derivatives equal to zero.

$$\frac{\partial f}{\partial x} = x^2 - 5 = 0, \qquad\qquad \frac{\partial f}{\partial y} = -6y^2 + 6 = 0,$$

$$x^2 = 5, \qquad\qquad 6y^2 = 6,$$

$$x = \sqrt{5}, \; x = -\sqrt{5}, \qquad\qquad y = 1, \; y = -1.$$

Thus we have four values (x,y) where there is a possible relative maximum or minimum:

$$(\sqrt{5},1), \quad (-\sqrt{5},1), \quad (\sqrt{5},-1), \quad (-\sqrt{5},-1).$$

13. $f(x,y) = 2x^2 - x^4 - y^2,$

$$\frac{\partial f}{\partial x} = 4x - 4x^3, \qquad\qquad \frac{\partial^2 f}{\partial x^2} = 4 - 12x^2,$$

$$\frac{\partial f}{\partial y} = -2y, \qquad\qquad \frac{\partial^2 f}{\partial y^2} = -2,$$

$$\frac{\partial^2 f}{\partial x \partial y} = 0.$$

$$D(x,y) = \frac{\partial^2 f}{\partial x^2} \cdot \frac{\partial^2 f}{\partial y^2} - \left(\frac{\partial^2 f}{\partial x \partial y}\right)^2$$

$$= (4 - 12x^2) \cdot (-2) - 0 = -2(4 - 12x^2).$$

Use the second-derivative test for functions of two variables. Looking initially at $(-1,0)$:

$$D(-1,0) = -2(4 - 12) \text{ which is positive; and}$$

$$\frac{\partial^2 f}{\partial x^2}(-1,0) = 4 - 12(-1)^2 \text{ which is negative.}$$

Hence $f(x,y)$ has a relative maximum at $(-1,0)$.

For $(0,0)$: $D(0,0) = -2(4 - 0) = -8$ which is negative, so $f(x,y)$ has neither a relative maximum nor a relative minimum at $(0,0)$.

Finally at $(1,0)$: $D(1,0) = -2(4 - 12)$ which is positive; and

$$\frac{\partial^2 f}{\partial x^2}(1,0) = 4 - 12 = -8 \text{ is negative.}$$

Therefore again by the test, $f(x,y)$ has a relative maximum at $(1,0)$.

19. $\frac{\partial}{\partial x}(-2x^2 + 2xy - y^2 + 4x - 6y + 5) = -4x + 2y + 4,$

$\frac{\partial}{\partial y}(-2x^2 + 2xy - y^2 + 4x - 6y + 5) = 2x - 2y - 6.$

Setting these partials equal to zero, we obtain

$$\frac{\partial f}{\partial x} = -4x + 2y + 4 = 0, \qquad\qquad \frac{\partial f}{\partial y} = 2x - 2y - 6 = 0,$$
$$2y = 4x - 4, \qquad\qquad\qquad\qquad 2y = 2x - 6,$$
$$y = 2x - 2, \qquad\qquad\qquad\qquad y = x - 3.$$

Equating these two expressions for y, we have

$$2x - 2 = x - 3,$$
$$x = -1.$$

Substituting $x = -1$ into $y = x - 3$, we obtain $y = x - 3 = (-1) - 3 = -4$. So there is a possible relative maximum or minimum at $(-1, -4)$. We use the second derivative test to determine which, if either, is the case.

$$\frac{\partial^2 f}{\partial x^2} = \frac{\partial}{\partial x}(-4x + 2y + 4) = -4,$$

$$\frac{\partial^2 f}{\partial y^2} = \frac{\partial}{\partial y}(2x - 2y - 6) = -2,$$

$$\frac{\partial^2 f}{\partial x \partial y} = \frac{\partial}{\partial x}(2x - 2y - 6) = 2.$$

Therefore, $D(x,y) = (-4)(-2) - (2)^2 = 8 - 4 = 4$, and hence $D(-1,-4) = 4$. Since $4 > 0$, $f(x,y)$ indeed has a relative maximum or minimum at $(-1,-4)$. Since $\frac{\partial^2 f}{\partial x^2}(-1,-4) = -4 < 0$, $f(x,y)$ has a relative maximum at $(-1,-4)$.

25. $\frac{\partial f}{\partial x} = 4x - 1, \quad \frac{\partial f}{\partial y} = 3y^2 - 12.$

Setting the two partials equal to 0 and solving, we obtain

$$4x - 1 = 0, \qquad\qquad\qquad 3y^2 - 12 = 0,$$
$$4x = 1, \qquad\qquad\qquad\qquad 3y^2 = 12,$$
$$x = 1/4, \qquad\qquad\qquad\qquad y^2 = 4,$$
$$\qquad\qquad\qquad\qquad\qquad\qquad y = 2, -2.$$

So there are two possible relative maxima and/or minima: $(1/4, 2)$, $(1/4, -2)$. For the second derivative test, we compute

$$\frac{\partial^2 f}{\partial x^2} = \frac{\partial}{\partial x}(4x - 1) = 4,$$

$$\frac{\partial^2 f}{\partial y^2} = \frac{\partial}{\partial y}(3y^2 - 12) = 6y,$$

$$\frac{\partial^2 f}{\partial x \partial y} = \frac{\partial}{\partial x}(3y^2 - 12) = 0.$$

$$D(x,y) = 4(6y) - 0^2 = 24y.$$

Now $D(1/4,2) = 24(2) = 48 > 0$, so $f(x,y)$ does indeed have a relative maximum or minimum at $(1/4,2)$. Since

$$\frac{\partial^2 f}{\partial x^2}(1/4,2) = 4 > 0,$$

there is a relative minimum at $(1/4,2)$. Finally, $D(1/4,-2) = 24(-2) = -48 < 0$, indicating that $(1/4,-2)$ is a saddle point; this is neither a relative maximum nor a relative minimum.

31. We note that the revenue can be expressed by $10x + 9y$, since the company sells x units of product I for 10 dollars each, and y units of product II for 9 dollars each. Noting also that profit = (revenue) − (cost), we write the profit function:

$$P(x,y) = 10x + 9y - [400 + 2x + 3y + .01(3x^2 + xy + 3y^2)]$$
$$= 8x + 6y - .03x^2 - .01xy - .03y^2 - 400.$$

We now proceed as in Exercises 19 and 25 above.

$$\frac{\partial P}{\partial x} = 8 - .06x - .01y,$$

$$\frac{\partial P}{\partial y} = 6 - .01x - .06y.$$

$8 - .06x - .01y = 0,$	$6 - .01x - .06y = 0,$
$.01y = 8 - .06x,$	$.06y = 6 - .01x,$
$y = 800 - 6x,$	$y = 100 - (1/6)x.$

So,
$$800 - 6x = 100 - (1/6)x,$$
$$700 = (35/6)x,$$
$$x = 120.$$

and
$$y = 800 - 6x = 800 - 6(120)$$
$$= 800 - 720 = 80.$$

So (120,80) is the only possible maximum. For the second derivative test, we compute

$$\frac{\partial^2 P}{\partial x^2} = -.06, \quad \frac{\partial^2 P}{\partial y^2} = -.06, \quad \frac{\partial^2 P}{\partial x \partial y} = -.01.$$

Therefore, $D(x,y) = (-.06)(-.06) - (-.01)^2 = .0035$. Thus $D(120,80) = .0035 > 0$ and (120,80) is indeed a maximum or minimum. Since $\frac{\partial^2 P}{\partial x^2}(120,80) = -.06 < 0$, (120,80) is a maximum. Thus profit is maximized by manufacturing and selling 120 units of product I and 80 units of product II.

7.4 Lagrange Multipliers and Constrained Optimization

Any letter of the alphabet can be used as a variable. In this section we introduce the Greek letter λ as a variable. The following results illustrate derivatives using λ.

$$\text{If } f(\lambda) = 5\lambda, \text{ then } \frac{df(\lambda)}{d\lambda} = 5.$$

$$\text{If } f(\lambda) = \lambda k, \text{ then } \frac{df(\lambda)}{d\lambda} = k.$$

Partial derivatives of functions involving λ are not difficult to compute once the unfamiliarity of using a Greek letter is overcome.

The problems in this section require that systems of three and four variables be solved. The method following Example 1 applies to all equations in x, y, and λ. Systems with the additional variable z can usually be solved by mimicking the solution to Example 4.

1. We construct the function

$$F(x,y,\lambda) = x^2 + 3y^2 + 10 + \lambda(8-x-y).$$

Setting the first partial derivatives equal to zero and solving for λ, we obtain

$$\frac{\partial F}{\partial x} = 2x - \lambda = 0, \text{ so } \lambda = 2x,$$

$$\frac{\partial F}{\partial y} = 6y - \lambda = 0, \text{ so } \lambda = 6y,$$

$$\frac{\partial F}{\partial \lambda} = 8 - x - y = 0.$$

Equating $\lambda = 2x$ and $\lambda = 6y$, we obtain $2x = 6y$, or $x = 3y$. Substituting 3y for x into the third equation, we have

$$8 - (3y) - y = 0,$$
$$8 - 4y = 0,$$
$$4y = 8,$$
$$y = 2.$$

Now $x = 3y = 3(2) = 6$ and $\lambda = 2x = 2(6) = 12$.

So the minimum is at $(6,2)$, and the minimum value is

$$F(6,2) = 6^2 + 3(2^2) + 10 = 36 + 12 + 10 = 58.$$

7. Minimize the function, $f(x,y) = x + y$ subject to the constraint

$$xy = 25, \text{ or } xy - 25 = 0.$$

$F(x,y,\lambda) = x + y + \lambda(xy - 25)$ is the Lagrange function. For a relative minimum we set its first partial derivatives equal to zero:

$$\left.\begin{array}{l} \dfrac{\partial F}{\partial F} = 1 + \lambda x = 0 \\[3ex] \dfrac{\partial F}{\partial y} = 1 + \lambda x = 0 \end{array}\right\} \Rightarrow \lambda y = \lambda x, \text{ or } y = x \quad (*),$$

$$\frac{\partial F}{\partial \lambda} = xy - 25 = 0 \text{ (substitute in } (*)),$$

$$x^2 - 25 = 0, \text{ or } x = \pm 5.$$

If $x = 5$ (since we need positive numbers): $y = 5$ by $(*)$. Therefore, $x = 5$, $y = 5$.

13. Maximize $P(x,y) = 3x + 4y$ subject to the constraint

$$9x^2 + 4y^2 - 18,000 = 0.$$

$P(x,y,\lambda) = 3x + 4y + \lambda(9x^2 + 4y^2 - 18,000)$. Setting the first partial derivatives equal to zero:

$$\frac{\partial p}{\partial x} = 3 + 18\lambda x = 0,$$

$$\frac{\partial p}{\partial y} = 4 + 8\lambda y = 0,$$

$$\frac{\partial p}{\partial \lambda} = 9x^2 + 4y^2 - 18000 = 0 \qquad (*).$$

Thus, $3 + 18\lambda x = 0$, and $4 + 8\lambda y = 0$, which gives:

$$1 + 6\lambda x = 0, \text{ and } 1 + 2\lambda y = 0, \text{ respectively.}$$

Therefore,
$$1 + 6\lambda x = 1 + 2\lambda y,$$

$$6\lambda x = 2\lambda y,$$
$$3x = y.$$

Substituting 3x for y in (*),

$$\text{i.e., } 9x^2 + 4(3x)^2 - 18,000 = 0,$$
$$45x^2 = 18,000,$$
$$x^2 = 400, \text{ or } x = \pm 20.$$

Since y = 3x, we have y = ±60. But x, y ≥ 0, so

$$x = 20, \text{ and } y = 60.$$

19. Let $F(x,y,z,\lambda) = 3x + 5y + z - x^2 - y^2 - z^2 + \lambda(6-x-y-z)$, and set the first partial derivatives equal to zero.

$$\frac{\partial F}{\partial x} = 3 - 2x - \lambda = 0,$$

$$\frac{\partial F}{\partial y} = 5 - 2y - \lambda = 0,$$

$$\frac{\partial F}{\partial z} = 1 - 2z - \lambda = 0.$$

This gives us three expressions for λ:

$$\lambda = 3 - 2x,$$
$$\lambda = 5 - 2y,$$
$$\lambda = 1 - 2z.$$

Equating the first two equations and solving for x, we have

$$3 - 2x = 5 - 2y,$$
$$-2x = 2 - 2y,$$
$$x = -1 + y.$$

Equating the second two equations and solving for z, we obtain:

$$5 - 2y = 1 - 2z,$$
$$2z = 2y - 4,$$
$$z = y - 2.$$

We now have expressions for x and for z in terms of y, and can substitute these into the equation

$$\frac{\partial F}{\partial \lambda} = 6 - x - y - z = 0.$$

$$6 - (-1+y) - y - (y-2) = 0$$
$$6 + 1 - y - y - y + 2 = 0$$
$$9 - 3y = 0$$

$$9 = 3y, \text{ or } y = 3.$$

Thus, $x = -1 + y = -1 + 3 = 2$ and $z = y - 2 = 3 - 2 = 1$. Thus the function is maximized when $x = 2$, $y = 3$, and $z = 1$.

25. Throughout the problem we are not given any specific production function, so we must use the general expression $f(x,y)$ in its place, which we wish to maximize. As a constraint we have that the cost of labor plus the cost of capital is c dollars, that is:

$$ax + by = c, \text{ or } ax + by - c = 0.$$

We then construct the function

$$F(x,y,\lambda) = f(x,y) + \lambda(ax + by - c).$$

We take partial derivatives and set them equal to zero.

$$\frac{\partial F}{\partial x} = \frac{\partial f}{\partial x} + \lambda a = 0,$$

$$\frac{\partial F}{\partial y} = \frac{\partial f}{\partial y} + \lambda b = 0.$$

Solving for λ we get

$$\lambda = \frac{-\frac{\partial f}{\partial x}}{a},$$

$$\lambda = \frac{-\frac{\partial f}{\partial y}}{b}.$$

And equating these expressions for λ, we conclude that

$$\frac{-\frac{\partial f}{\partial x}}{a} = \frac{-\frac{\partial f}{\partial x}}{b},$$

$$-b\left(\frac{\partial f}{\partial x}\right) = -a\left(\frac{\partial f}{\partial y}\right),$$

$$\frac{\frac{\partial f}{\partial x}}{\frac{\partial f}{\partial y}} = \frac{-a}{-b} = \frac{a}{b}.$$

7.5 The Method of Least Squares

Each problem in this section requires the solution of a system of two equations in two unknowns. These systems differ from those previously studied in three ways.

1. The variables (that is, the unknowns) will always be A and B, rather than x and y.

2. The equations are linear. That is, they have the form aA + bB = c, where a, b, and c are constants.

3. The systems have exactly one solution. This solution is best found by multiplying one equation by a constant and subtracting it from the other equation to eliminate one of the variables. The value of the other variable is then easily found. By substituting this value into one of the equations, the value of the first variable is easily found.

1. Let the straight line be $y = Ax + B$. When $x = 1, 2, 3$ the corresponding y-coordinates of the points of the line are $A+B$, $2A+B$, $3A+B$. Therefore, the squares of the vertical distances from the line to the points $(1,0)$, $(2,-1)$, and $(3,-4)$ are

$$E_1^2 = (A+B-0)^2 = (A+B)^2,$$

$$E_2^2 = (2A+B-(-1))^2 = (2A+B+1)^2,$$

$$E_3^2 = (3A+B-(-4))^2 = (3A+B+4)^2.$$

The least-squares error is

$$f(A,B) = E_1^2 + E_2^2 + E_3^2 = (A+B)^2 + (2A+B+1)^2 + (3A+B+4)^2.$$

To minimize $f(A,B)$, we take partial derivatives with respect to A and B, and set them equal to zero.

$$\frac{\partial f}{\partial A} = 2(A+B) + 2(2A+B+1)\cdot 2 + 2(3A+B+4)\cdot 3$$

$$= (2A+2B) + (8A+4B+4) + (18A+6B+24)$$

$$= 28A + 12B + 28 = 0,$$

$$\frac{\partial f}{\partial B} = 2(A+B) + 2(2A+B+1) + 2(3A+B+4)$$

$$= (2A+2B) + (4A+2B+2) + (6A+2B+8)$$

$$= 12A + 6B + 10 = 0.$$

To solve for A and B we must solve the system of simultaneous linear equations

$$28A + 12B = -28,$$
$$12A + 6B = -10.$$

Multiplying the second equation by 2 and subtracting from the first equation we have

$$4A = -8 \text{ or } A = -2.$$

Substituting $A = -2$ into the second linear equation,

$$12(-2) + 6B = -10,$$
$$6B = 14,$$
$$B = 14/6 = 7/3.$$

The straight line that minimizes the least-squares error is $y = -2x + 7/3$.

Helpful Hint: Here are two tips on solving the systems of linear equations.

1. Twos can be cancelled out of each equation at an early stage. For instance, a simpler form of the first equation in Exercise 1 can be obtained as follows.

$$\frac{1}{2} \cdot \frac{\partial f}{\partial A} = 2(A+B) + 2(2A+B+1) \cdot 2 + 2(3A+B+4) \cdot 3 = 0,$$

$$14A + 6B + 14 = 0.$$

2. The coefficient of B in the second equation will always be a whole number. If the twos are factored out as in the equation above, this value will equal the number of data points. Therefore, B will usually be the best variable to eliminate when solving the system of equations.

These tips are used in the solution to Exercise 7.

7. Let the straight line be $y = Ax + B$. When $x = 1,2,4,5$ the corresponding y-coordinates of the points on the line are $A + B$, $2A + B$, $4A + B$, $5A + B$. Therefore, the squares of the vertical distances from the line to the points $(1,0)$, $(2,1)$, $(4,2)$, $(5,3)$ are:

$$E_1^2 = (A+B-0)^2 = (A+B)^2,$$

$$E_2^2 = (2A+B-1)^2,$$

$$E_3^2 = (4A+B-2)^2,$$

$$E_4^2 = (5A+B-3)^2.$$

Thus, the least-squares error is

$$f(A,B) = E_1^2 + E_2^2 + E_3^2 + E_4^2 = (A+B)^2 + (2A+B-1)^2$$

$$+ (4A+B-2)^2 + (5A+B-3)^2.$$

To minimize $f(A,B)$, we take partial derivatives with respect to A and B, and set them equal to zero.

$$\frac{\partial f}{\partial A} = 2(A+B) + 2(2A+B-1) \cdot 2$$

$$+ 2(4A+B-2) \cdot 4 + 2(5A+B-3) \cdot 5 = 0.$$

Dividing each term by 2, we obtain

$$46A + 12B - 25 = 0.$$

Next,

$$\frac{\partial f}{\partial B} = 2(A+B) + 2(2A+B-1) + 2(4A+B-2) + 2(5A+B-3) = 0.$$

Dividing each term by 2, we obtain

$$12A + 4B - 6 = 0.$$

To solve for A and B, we must solve the system of simultaneous linear equations

$$\begin{cases} 46A + 12B = 25, \\ 12A + 4B = 6. \end{cases}$$

Multiplying the second equation by 3 and subtracting from the first equation we have

$$10A = 7, \text{ or } A = \frac{7}{10} = .7.$$

Since $12A + 4B = 6$ we have

$$4B = 6 - 12(.7) = 6 - 8.4 = -2.4.$$

Hence $B = \frac{-2.4}{4} = -.6$, and the straight line that minimizes the least-squares error is $y = .7x - .6$.

13. We make a prediction by determining the line $y = Ax + B$ that best fits these data using the method of least squares. We then set $x = 7$ and find the corresponding y value on this line. When $x = 1,2,3,4,5$, the corresponding y-coordinates of the points on the line are A+B, 2A+B, 3A+B, 4A+B, and 5A+B. Therefore, the squares of the vertical distances from the line to the points (1,38), (2,40), (3,41), (4,39), (5,45) are

$$E_1^2 = (A+B-38)^2, \qquad\qquad E_2^2 = (2A+B-40)^2,$$

$$E_3^2 = (3A+B-41)^2, \qquad\qquad E_4^2 = (4A+B-39)^2,$$

$$E_5^2 = (5A+B-45)^2.$$

Thus, the least-squares error is:

$$f(A,B) = E_1^2 + E_2^2 + E_3^2 + E_4^2 + E_5^2$$

$$= (A+B-38)^2 + (2A+B-40)^2 + (3A+B-41)^2$$

$$+ (4A+B-39)^2 + (5A+B-45)^2.$$

To minimize $f(A,B)$, we take partial derivatives with respect to A and B, and set them equal to zero.

$$\frac{\partial f}{\partial A} = 2(A+B-38) + 2(2A+B-40)\cdot 2 + 2(3A+B-41)\cdot 3$$

$$+ 2(4A+B-39)\cdot 4 + 2(5A+B-45)\cdot 5.$$

Setting $\frac{\partial f}{\partial A} = 0$ and dividing each term in the equation by 2, we have

$$(A+B-38) + (4A+2B-80) + (9A+3B-123)$$

$$+ (16A+4B-156) + (25A+5B-225) = 0$$

and

$$55A + 15B = 622. \tag{1}$$

Next,

$$\frac{\partial f}{\partial B} = 2(A+B-38) + 2(2A+B-40) + 2(3A+B-41)$$

$$+ 2(4A+B-39) + 2(5A+B-45).$$

Setting $\frac{\partial f}{\partial B} = 0$ and dividing each term in the equation by 2, we have

$$(A+B-38) + (2A+B-40) + (3A+B-41)$$

$$+ (4A+B-39) + (5A+B-45) = 0$$

and

$$15A + 5B = 203. \tag{2}$$

To solve for A and B, we must solve the system of simultaneous linear equations (1) and (2). Multiplying the equation in (2) by −3 and adding to the equation in (1), we have

$$10A = 13, \text{ or } A = \frac{13}{10} = 1.3.$$

Substituting this into (2) we have

$$15(1.3) + 5B = 203,$$
$$5B = 203 - 19.5 = 183.5.$$

Hence B = 36.7 and the line that minimizes the least-squares error is y = 1.3x + 36.7. We use this line to estimate the number of cars sold during week 7 by setting x = 7. Then we get y = 1.3(7) + 36.7 = 9.1 + 36.7 = 45.8. Since only a whole number of cars can be sold, we round this off and estimate that 46 cars will be sold during the seventh week.

7.6 Double Integrals

The problems in this section amount to performing two integrations, once with respect to y as the variable and once with respect to x. Antidifferentiation with respect to y takes a little getting used to. It is just the reverse operation to taking partial derivatives with respect to y.

1. We first evaluate the inner integral.

$$\int_0^1 e^{x+y} \, dy = e^{x+y} \Big|_0^1 = e^{x+1} - e^x.$$

[The dy in the integral tells us that y is the variable of integration in $e^{x+y} \Big|_0^1$.] We can now evaluate the outer integral.

$$\int_0^1 (e^{x+1} - e^x) \, dx = (e^{x+1} - e^x) \Big|_0^1$$
$$= (e^2 - e) - (e - 1)$$
$$= e^2 - 2e + 1.$$

7. We first evaluate the inner integral.

$$\int_x^{2x} (x+y) \, dy = (xy + \tfrac{1}{2}y^2) \Big|_x^{2x}$$
$$= [x(2x) + \tfrac{1}{2}(2x)^2] - [x \cdot x + \tfrac{1}{2}x^2]$$
$$= [2x^2 + 2x^2] - x^2 - \tfrac{1}{2}x^2$$
$$= \tfrac{5}{2}x^2.$$

We can now evaluate the outer integral.

$$\int_1^1 \tfrac{5}{2}x^2 \, dx = \tfrac{5}{6}x^3 \Big|_{-1}^{1} = \tfrac{5}{6} - (-\tfrac{5}{6}) = \tfrac{5}{3} \; .$$

13. The desired volume is given by the double integral:

$$\iint_R (x^2+y^2) \, dx \, dy,$$

which is equivalent to the iterated integral:

$$\int_1^3 \left(\int_0^1 (x^2+y^2) \, dy \right) dx.$$

Evaluating the inner integral we have

$$\int_0^1 (x^2+y^2) \, dy = (x^2 y + \tfrac{1}{3}y^3) \Big|_0^1$$

$$= (x^2 + \tfrac{1}{3}) - (0 + 0) = x^2 + \tfrac{1}{3} \; .$$

Evaluating the outer integral we have

$$\int_1^3 (x^2 + \tfrac{1}{3}) \, dx = (\tfrac{1}{3}x^3 + \tfrac{1}{3}x) \Big|_1^3 = (9+1) - (\tfrac{1}{3}+\tfrac{1}{3}) = 9\tfrac{1}{3} \; .$$

Chapter 7: Supplementary Exercises

1. $f(2,9) = \dfrac{2\sqrt{9}}{1+2} = \dfrac{6}{3} = 2,$

    $f(5,1) = \dfrac{5\sqrt{1}}{1+5} = \dfrac{5}{6},$

    $f(0,0) = \dfrac{0\sqrt{0}}{1+0} = \dfrac{0}{1} = 0.$

7. $f(x,y) = e^{x/y} = e^{xy^{-1}},$

    $\dfrac{\partial f}{\partial x} = y^{-1}e^{xy^{-1}} = \dfrac{1}{y}e^{x/y}$, (we treat y as a constant)

    $\dfrac{\partial f}{\partial y} = e^{xy^{-1}}[\dfrac{\partial}{\partial y}xy^{-1}] = -xy^{-2}e^{xy^{-1}} = -\dfrac{x}{y^2}e^{x/y}$, (we treat x as a constant).

13. $f(x,y) = x^5 - 2x^3y + \frac{1}{2}y^4$.

In order to find the second partial derivatives we must first determine the first partial derivatives with respect to x and y.

$$\frac{\partial f}{\partial x} = 5x^4 - 6x^2y,$$

$$\frac{\partial f}{\partial y} = -2x^3 + 2y^3.$$

Differentiating each of these with respect to x and y, we have

$$\frac{\partial^2 f}{\partial x^2} = \frac{\partial}{\partial x}\left(\frac{\partial f}{\partial x}\right) = \frac{\partial}{\partial x}(5x^4 - 6x^2y) = 20x^3 - 12xy,$$

$$\frac{\partial^2 f}{\partial y^2} = \frac{\partial}{\partial y}\left(\frac{\partial f}{\partial y}\right) = \frac{\partial}{\partial y}(-2x^3 + 2y^3) = 6y^2,$$

$$\frac{\partial^2 f}{\partial x \partial y} = \frac{\partial}{\partial x}\left(\frac{\partial f}{\partial y}\right) = \frac{\partial}{\partial x}(-2x^3 + 2y^3) = -6x^2,$$

$$\frac{\partial^2 f}{\partial y \partial x} = \frac{\partial}{\partial y}\left(\frac{\partial f}{\partial x}\right) = \frac{\partial}{\partial y}(5x^4 - 6x^2y) = -6x^2.$$

19. $\frac{\partial}{\partial x}(x^3 + 3x^2 + 3y^2 - 6y + 7) = 3x^2 + 6x,$

$\frac{\partial}{\partial y}(x^3 + 3x^2 + 3y^2 - 6y + 7) = 6y - 6.$

To find possible relative maxima or minima, set the first partial derivatives equal to zero.

$$\frac{\partial f}{\partial x} = 3x^2 + 6x = 0, \qquad\qquad \frac{\partial f}{\partial y} = 6y - 6 = 0,$$

$$3x(x+2) = 0, \qquad\qquad\qquad 6y = 6,$$

$$x = 0, \ x = -2, \qquad\qquad\qquad y = 1.$$

Thus $f(x,y)$ has a possible relative maximum or minimum at $(0,1)$ and at $(-2,1)$.

25. We construct the function

$$F(x,y,\lambda) = 3x^2 + 2xy - y^2 + \lambda(5-2x-y).$$

Setting the first partial derivatives equal to zero and solving for λ we obtain

$$\frac{\partial F}{\partial x} = 6x + 2y - 2\lambda = 0,$$

$$2\lambda = 6x + 2y,$$

$$\lambda = 3x + y.$$

$$\frac{\partial F}{\partial y} = 2x - 2y - \lambda = 0,$$

$$\lambda = 2x - 2y.$$

Equating the two expressions for λ we have

$$3x + y = 2x - 2y \quad \text{or} \quad x = -3y.$$

We have $\frac{\partial F}{\partial \lambda} = 5 - 2x - y = 0$ as a constraint, and substituting $-3y$ for x giving:

$$5 - 2(-3y) - y = 0,$$

$$5 + 5y = 0,$$

$$5y = -5, \text{ or } y = -1.$$

Hence, $x = -3y = -3(-1) = 3$, $\lambda = 2x - 2y = 2(3) - 2(-1) = 6 + 2 = 8$, so the maximum is at $(3,-1)$, and the maximum value is $f(3,-1) = 3(9) + 2(3)(-1) - 1 = 27 - 6 - 1 = 20$.

31. Let the straight line be $y = Ax + B$. When $x = 1,2,3$ the corresponding y-coordinates of the points of the line are $A+B$, $2A+B$, $3A+B$, respectively. Therefore, the squares of the vertical distances from the line to the points $(1,1)$, $(2,3)$, and $(3,6)$ are

$$E_1^2 = (A+B - 1)^2,$$

$$E_2^2 = (2A+B - 3)^2,$$

$$E_3^2 = (3A+B - 6)^2.$$

Thus the least-squares error is

$$f(A,B) = E_1^2 + E_2^2 + E_3^2$$

$$= (A + B - 1)^2 + (2A + B - 3)^2 + (3A + B - 6)^2.$$

To minimize $f(A,B)$, we take partial derivatives with respect to A and B, and set them equal to zero.

$$\frac{\partial f}{\partial A} = 2(A + B - 1) + 2(2A + B - 3) \cdot 2 + 2(3A + B - 6) \cdot 3$$

Don't forget these factors ⟶ ↑   ↑

$$= (2A + 2B - 2) + (8A + 4B - 12) + (18A + 6B - 36)$$

$$= 28A + 12B - 50 = 0.$$

$$\frac{\partial f}{\partial B} = 2(A + B - 1) + 2(2A + B - 3) + 2(3A + B - 6)$$

$$= (2A + 2B - 2) + (4A + 2B - 6) + (6A + 2B - 12)$$

$$= 12A + 6B - 20 = 0.$$

To solve for A and B, we must solve the system of simultaneous linear equations

$$28A + 12B = 50,$$
$$12A + 6B = 20.$$

Subtracting 2 times the second equation from the first equation we have

$$4A = 10, \text{ and } A = 5/2.$$

Hence

$$6B = 20 - 12(5/2) = -10, \text{ and } B = -5/3.$$

Therefore, the straight line that minimizes the least-squares error is $y = (5/2)x - 5/3$.

37. $\iint\limits_R 5 \, dx \, dy$ represents the volume of a box with dimensions

$(4 - 0) \times (3 - 1) \times 5$. So $\iint\limits_R 5 \, dx \, dy = 4 \cdot 2 \cdot 5 = 40.$

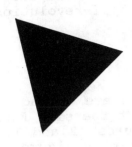

CHAPTER 8

THE TRIGONOMETRIC FUNCTIONS

Trigonometric functions are useful for describing physical proceses that are cyclical or periodic. The goal of this short chapter is to study basic properties of these functions. We realize that many of our readers have never studied trigonometry, and so we mention a few simple connections with right triangles. However, our main interest is in calculus, not in trigonometry.

8.1 Radian Measure of Angles

All of the exercises in this text use radian measure in connection with trigonometric functions.

1. $30° = \overset{}{\cancel{30}} \times \dfrac{\pi}{\cancel{180}_{6}} \text{ radians} = \dfrac{\pi}{6} \text{ radians}$

    $120° = \overset{2}{\cancel{120}} \times \dfrac{\pi}{\cancel{180}_{3}} \text{ radians} = \dfrac{2\pi}{3} \text{ radians}$

    $315° = \overset{7}{\cancel{315}} \times \dfrac{\pi}{\cancel{180}_{4}} \text{ radians} = \dfrac{7\pi}{4} \text{ radians.}$

7. The angle described by this figure consists of one full revolution plus three quarters of a revolution. That is

    $$t = 2\pi + \frac{3}{4}(2\pi) = 2\pi + \frac{3}{2}\pi = \frac{7\pi}{2}.$$

13. See the figures in the answer section of the text. Here is the reasoning that leads to those figures. First, since $\pi/2$ is one quarter-revolution of the circle, $3\pi/2$ is three quarter-revolutions. Next, since $\pi/4$ is one eighth-revolution of the circle, $3\pi/4$ is three eighth-revolutions. Notice also that $3\pi/4$ is exactly one half the angle de-

scribed in the first case. Finally, since π is one half-revolution of the circle, 5π is five half-revolutions or two-and-one-half revolutions.

8.2 The Sine and the Cosine

The most important pictures in this section are on pages 519-20. Years from now if someone asks you what the sine function is, you should think of the graphs in Figures 12 and 13, not the triangles in Figure 3, or even the circle in Figure 7. It is the regular, fluctuating shape of the graphs of $\cos t$ and $\sin t$ that make them useful in mathematical models. Nevertheless, your instructor will want you to have a basic understanding of how $\cos t$ and $\sin t$ are defined, and that is the purpose of the other figures in this section.

Exercises 1—20 help you learn the basic definitions:

$$\cos t = \frac{x}{r}, \quad \text{and} \quad \sin t = \frac{t}{r},$$

where x and y are the coordinates of a point that determines an angle of t radians, and r is the distance of the point from the origin. Exercises 21-38 help you learn the alternative definitions of $\cos t$ and $\sin t$ as the x- and y-coordinates of an appropriate point on the unit circle. The alternative definitions enable you to understand some important properties of the sine and cosine. Be sure to find out which of these properties you must memorize. [Formulas (7) and (8), for example, are only used later for proofs of derivative formulas, so you may not need to learn them.]

1. $\sin t = \dfrac{\text{opposite}}{\text{hypotenuse}} = \dfrac{1}{2}$, $\cos t = \dfrac{\text{adjacent}}{\text{hypotenuse}} = \dfrac{\sqrt{3}}{2}$.

7. First we must compute r.

$$r = \sqrt{x^2+y^2} = \sqrt{(-2)^2+1^2} = \sqrt{4+1} = \sqrt{5}.$$

Then

$$\sin t = \frac{y}{r} = \frac{1}{\sqrt{5}}, \quad \cos t = \frac{x}{r} = \frac{-2}{\sqrt{5}}.$$

13. First compute

$$\sin t = \frac{b}{c} = \frac{5}{13} = .385.$$

One way to find t is to look in the "$\sin t$" column of a table of trigonometric functions. A number in this column close to .385 is .38942, in the row corresponding to t = .4. From this you can conclude that $t \approx .4$. For more accuracy, use a scientific calculator. Check to make sure that angle measurement is set for radians, and use the "inverse sine" function to compute $\sin^{-1} .385 = .39521$.

19. Since we know "a" and it is the side <u>adjacent</u> to the given angle t, we use the formula

$$\cos t = \frac{\text{adjacent}}{\text{hypotenuse}}.$$

We have

$$\cos .5 = \frac{2.4}{c}.$$

From a calculator, $\cos .5 = .87758$, so

$$.87758 = \frac{2.4}{c},$$

$$.87758c = 2.4,$$

$$c = \frac{2.4}{.87758} = 2.7348.$$

One way to find b, now that we know c, is to use the formula

$$\sin t = \frac{b}{c}.$$

That is,

$$\sin .5 = \frac{b}{2.73},$$

$$.47943 = \frac{b}{2.73},$$

$$2.73(.47943) = b,$$

$$b = 1.31.$$

Alternately, using the Pythagorean theorem, we have

$$a^2 + b^2 = c^2, \text{ i.e., } (2.4)^2 + b^2 = (2.73)^2,$$

$$5.76 + b^2 = 7.45,$$

$$b^2 = 1.69,$$

$$b = 1.30.$$

The discrepancy between the two values for b is due to the round-off error in using 2.73 in place of 2.7348. Either value of b is accurate enough for our purposes.

25. On the unit circle we locate the point P that is determined by an angle of $\frac{-5\pi}{8}$ radians. The x-coordinate of P is $\cos\left(\frac{-5\pi}{8}\right)$. There is another point Q on the unit circle

with the same x-coordinate. (See the figure.)   Let t be the radian measure of the angle determined by Q.   Then $\cos t = \cos(\frac{-5\pi}{8})$ because Q and P have the same x-coordinate. Also, $0 \le t \le \pi$. From the symmetry of the diagram, it is clear that $t = \frac{5\pi}{8}$.   Also, since $\cos(-t) = \cos t$ and $0 \le \frac{5\pi}{8} \le \pi$, we have $t = \frac{5\pi}{8}$.

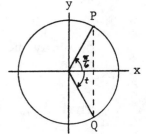

31. The equation $\sin t = -\sin(\pi/6)$ relates the y-coordinates of two points on the unit circle. One point P corresponds to an angle of $\pi/6$ radians. (See the figure.) The other point Q must be on the right half of the unit circle, since $-\frac{\pi}{6} \le t \le \frac{\pi}{2}$. But its y-coordinate must be the negative of the y-coordinate of P.   We conclude from the figure that $t = -\frac{\pi}{6}$.

Another method is to recall that $-\sin t = \sin(-t)$.   Thus

$$-\sin(\pi/6) = \sin(-\pi/6).$$

Since $-\pi/6$ is between $-\pi/2$ and $\pi/2$, we conclude that $t = -\pi/6$.

37. Thinking of $\sin t$ as the y-coordinate of the point on the unit circle that is determined by an angle of t radians, we find that $\sin 5\pi = 0$ and $\sin(-2\pi) = 0$.

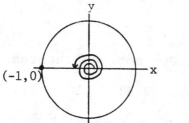
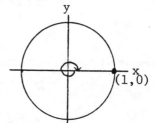

Since $\frac{17\pi}{2} = 8\pi + \frac{\pi}{2} = 4(2\pi) + \frac{\pi}{2}$, $\frac{17\pi}{2}$ is four complete revolutions of the circle plus one quarter-revolution. So P is the point $(0,1)$ and $\sin\frac{17\pi}{2} = 1$.

Since $-\frac{13\pi}{2} = -6\pi - \frac{\pi}{2} = 3(-2\pi) - \frac{\pi}{2}$, $-\frac{13\pi}{2}$ is three complete revolutions of the circle in the negative direction plus one quarter-revolution in the negative direction. So P is the point $(0,-1)$ and $\sin(-\frac{13\pi}{2}) = -1$.

8.3 Differentiation of sin t and cos t

The significance of the derivative formula

$$\frac{d}{dt} \sin t = \cos t$$

is revealed by Figures 1 and 2 in the text. That is, for each number t, the value of cos t gives the slope of the sine curve at the specified value of t. This fundamental property of the derivative as a slope function is worth contemplating because it will be crucial for our work in Chapter 10.

Most of the exercises in this section are routine drill problems. Exercises 35-38 are just a "warm-up" to remind you of antiderivatives. We will concentrate on problems such as these in Section 9.1.

Helpful Hint: <u>Don't</u> memorize antiderivative formulas for cos t and sin t, because you might confuse them with the derivative formulas. Just memorize where the minus sign goes in the derivative formulas, and then use this knowledge to check your work whenever you need to "guess" an antiderivative involving a sine or cosine. (See our solution to Exercise 37, below.)

1. By the chain rule we have

$$\frac{d}{dt}(\sin 4t) = (\cos 4t)\cdot\frac{d}{dt}4t = 4 \cos 4t.$$

7. $\frac{d}{dt}(t + \cos \pi t) = \frac{d}{dt}t + \frac{d}{dt}\cos \pi t$

$$= 1 + (-\sin \pi t)\frac{d}{dt}\pi t = 1 - \pi \sin \pi t.$$

13. By repeated application of the chain rule we have

$$\frac{d}{dx} \sin \sqrt{x-1} = \cos \sqrt{x-1}\cdot\frac{d}{dx}\sqrt{x-1}$$

$$= \cos \sqrt{x-1}\cdot\frac{d}{dx}(x-1)^{1/2}$$

$$= \cos \sqrt{x-1}\cdot(\frac{1}{2}(x-1)^{-1/2})\frac{d}{dx}(x-1)$$

$$= \cos \sqrt{x-1}\cdot(\frac{1}{2}(x-1)^{-1/2})$$

$$= \frac{\cos \sqrt{x-1}}{2\sqrt{x-1}}.$$

19. By repeated application of the chain rule we have

$$\frac{d}{dx}\cos^2 x^3 = \frac{d}{dx}(\cos x^3)^2$$

$$= 2\cos x^3 \cdot \frac{d}{dx}\cos x^3$$

$$= 2\cos x^3(-\sin x^3)\cdot\frac{d}{dx}x^3$$

$$= 2\cos x^3(-\sin x^3)3x^2$$

$$= -6x^2\cos x^3 \sin x^3.$$

Helpful Hint: Instructors like problems like Exercise 19 because it forces students to understand the notation. To avoid mistakes, be sure to include the first step shown in the solution above.

25. By the quotient rule we have

$$\frac{d}{dt}\left(\frac{\sin t}{\cos t}\right) = \frac{(\cos t)\cdot\frac{d}{dt}\sin t - (\sin t)\cdot\frac{d}{dt}\cos t}{\cos^2 t}$$

$$= \frac{\cos^2 t - (-\sin^2 t)}{\cos^2 t} = \frac{\cos^2 t + \sin^2 t}{\cos^2 t}$$

$$= \frac{1}{\cos^2 t} \quad [\text{because } \cos^2 t + \sin^2 t = 1].$$

An alternate notation for the answer is $\cos^{-2} t$.

31. $y' = -3\sin 3x$

So at $x = \frac{13\pi}{6}$ the slope of the tangent line is

$$-3\sin 3(\tfrac{13\pi}{6}) = -3\sin\frac{13\pi}{6} = -3\sin(6\pi + \tfrac{\pi}{2}) = -3\sin\frac{\pi}{2}$$
$$= -3\cdot 1 = -3.$$

37. Our first guess is that an antiderivative of $\sin(4x+1)$ should involve the cosine function, namely, $\cos(4x+1)$. We may differentiate our guess (<u>not</u> the original function) to check our reasoning:

$$\frac{d}{dx}\cos(4x+1) = -\sin(4x+1)\cdot 4 = -4\sin(4x+1).$$

This derivative is -4 times what it should be. So we adjust our guess by dividing it by -4. Now

$$\frac{d}{dx}\left[-\frac{1}{4}\cos(4x+1)\right] = -\frac{1}{4}\cdot\frac{d}{dx}\cos(4x+1)$$

$$= -\frac{1}{4}\left[-\sin(4x+1)\cdot 4\right] = \sin(4x+1).$$

Thus $-\frac{1}{4}\cos(4x+1)$ is an antiderivative of $\sin(4x+1)$. Hence

$$\int \sin(4x+1)\,dx = -\frac{1}{4}\cos(4x+1) + C.$$

43. Notice that $\displaystyle \lim_{h\to 0}\frac{\sin\left(\frac{\pi}{2}+h\right)-1}{h} = \lim_{h\to 0}\frac{\sin\left(\frac{\pi}{2}+h\right)-\sin\frac{\pi}{2}}{h}.$

Recall from Chapter 1 that difference quotients of the form $\frac{f(a+h)-f(a)}{h}$ are used to approximate the derivative $f'(a)$. In this exercise, $f(x)=\sin x$ and $a=\pi/2$. So the limit above is just the derivative of $\sin x$ at $\pi/2$. Since $(\sin x)'=\cos x$, and $\cos(\pi/2)=0$, the limit is 0.

8.4 The Tangent and Other Trigonometric Functions

You may need to memorize the discussion of how to get a derivative formula for $\tan t$. (See the bottom of page 539 or the Study Guide solution to Exercise 25 in Section 8.2.) This is a nice application of the quotient rule. Of course you should also memorize the formula:

$$\boxed{\frac{d}{dt}\tan t = \sec^2 t.}$$

1. Since $\cos t = \dfrac{\text{adjacent}}{\text{hypotenuse}}$ and $\sec t = \dfrac{1}{\cos t}$, we have

$$\sec t = \frac{\text{hypotenuse}}{\text{adjacent}}.$$

7. $\tan t = \dfrac{y}{x} = \dfrac{2}{-2} = -1.$ To find $\sec t$ we need r.

$$r = \sqrt{x^2+y^2} = \sqrt{(-2)^2+2^2} = \sqrt{4+4} = \sqrt{8} = 2\sqrt{2},$$

$$\sec t = \frac{r}{x} = \frac{2\sqrt{2}}{-2} = -\sqrt{2}.$$

13. Note that $\sec t = 1/\cos t = (\cos t)^{-1}$. Hence

$$\frac{d}{dt}\sec t = \frac{d}{dt}(\cos t)^{-1} = (-1)(\cos t)^{-2}\cdot\frac{d}{dt}\cos t$$

$$= (-1)(\cos t)^{-2}(-\sin t) = \frac{\sin t}{(\cos t)^2}.$$

This answer is acceptable. It may be written in other equivalent forms, such as $\sin t \sec^2 t$, or

$$\frac{\sin t}{\cos t} \cdot \frac{1}{\cos t} = \tan t \sec t.$$

19. By the chain rule,

$$f'(x) = \frac{d}{dx} 3\tan(\pi-x)$$

$$= 3\sec^2(\pi-x) \cdot \frac{d}{dx}(\pi-x)$$

$$= -3\sec^2(\pi-x).$$

25. By the product rule,

$$y' = \frac{d}{dx}(x \tan x)$$

$$= x \cdot \frac{d}{dx}(\tan x) + (\tan x) \cdot \frac{d}{dx} x$$

$$= x \sec^2 x + \tan x.$$

31. Using the chain rule, and the solution to Exercise 13, we have

$$y' = \frac{d}{dt} \ln(\tan t + \sec t)$$

$$= \frac{1}{\tan t + \sec t} \frac{d}{dt}(\tan t + \sec t)$$

$$= \frac{1}{\tan t + \sec t}\left(\frac{d}{dt}\tan t + \frac{d}{dt}\sec t\right)$$

$$= \frac{1}{\tan t + \sec t}(\sec^2 t + \tan t \sec t)$$

$$= \frac{(\sec t + \tan t)\sec t}{\tan t + \sec t} = \sec t.$$

Chapter 8: Supplementary Exercises

Use the chapter checklist to review the main definitions and facts. You may also need to memorize some of the following facts: (1) the graphs of $\sin t$ and $\cos t$ (knowing the values of $\sin t$ and $\cos t$ when $t = 0$, $\pi/2$, π, $3\pi/2$, and 2π, for example); (2) selected trig identities from Section 8.2; and (3) the definitions of $\sec t$, $\cot t$, and $\csc t$. Check with your instructor.

1. The angle described is three quarter-revolutions of the circle or $3(\frac{\pi}{2}$ radians$) = \frac{3\pi}{2}$ radians.

7. First we find r.
$$r = \sqrt{x^2+y^2} = \sqrt{3^2+4^2} = \sqrt{25} = 5.$$
Therefore,
$$\sin t = \frac{y}{r} = \frac{4}{5} = .8, \quad \cos t = \frac{x}{r} = \frac{3}{5} = .6, \quad \tan t = \frac{y}{x} = \frac{4}{3}.$$

13. $\sin t = \cos t$ whenever the point on the unit circle defined by the angle t has identical x and y coordinates (i.e., the point lies on the line $x = y$). The points on the unit circle with this property are $(1/\sqrt{2},\ 1/\sqrt{2})$ and $(-1/\sqrt{2}, -1/\sqrt{2})$. The four angles between -2π and 2π that correspond to these points are $-\frac{3\pi}{4}, -\frac{7\pi}{4}, \frac{\pi}{4}, \frac{5\pi}{4}$.

19. $f'(t) = \frac{d}{dt} 3\sin t = 3\cos t.$

25. By the quotient rule,
$$f'(x) = \frac{d}{dx}(\frac{\cos 2x}{\sin 3x})$$
$$= \frac{\sin(3x)\cdot\frac{d}{dx}\cos 2x - \cos(2x)\cdot\frac{d}{dx}\sin 3x}{(\sin 3x)^2}$$
$$= \frac{\sin(3x)(-2\sin 2x) - 3\cos(2x)\cos(3x)}{\sin^2(3x)}$$
$$= -\frac{2\sin(3x)\sin(2x) + 3\cos(2x)\cos(3x)}{\sin^2(3x)}.$$

31. By the chain rule,
$$y' = \frac{d}{dx}\sin(\tan x) = \cos(\tan x)\frac{d}{dx}\tan x$$
$$= \cos(\tan x)\sec^2 x.$$

37. By the product and chain rules,
$$y' = \frac{d}{dx} e^{3x}\sin^4 x$$
$$= e^{3x}(4\sin^3 x)\frac{d}{dx}\sin x + (\sin^4 x)3e^{3x}$$
$$= 4e^{3x}\sin^3 x \cos x + 3e^{3x}\sin^4 x.$$

43. $f(t) = \sin^2 t$.

$$f'(t) = \frac{d}{dt}(\sin t)^2$$

$$= (2\sin t)\cdot\frac{d}{dx}\sin t$$

$$= 2\sin t\cdot\cos t.$$

$$f''(t) = \frac{d}{dt}(2\sin t\cdot\cos t)$$

$$= 2\left[(\sin t)\cdot\frac{d}{dt}\cos t + (\cos t)\cdot\frac{d}{dt}\sin t\right]$$

$$= 2[(\sin t)(-\sin t) + \cos t\cdot\cos t]$$

$$= 2(\cos^2 t - \sin^2 t).$$

49. $y' = \frac{d}{dt}\tan t = \sec^2 t$.

The slope of tangent line when $t = \frac{\pi}{4}$ is

$$\sec^2(\frac{\pi}{4}) = \left(\frac{1}{\cos\frac{\pi}{4}}\right)^2 = \left(\frac{1}{\frac{\sqrt{2}}{2}}\right)^2 = 2.$$

At $t = \frac{\pi}{4}$, $y = \tan\frac{\pi}{4} = 1$, so $(\frac{\pi}{4}, 1)$ is a point on the line. Hence the equation of the line is

$$y - 1 = 2(t - \frac{\pi}{4}).$$

55. It is easy to check that the line $y = x$ is tangent to the graph of $y = \sin x$ at $x = 0$. From the graph of $y = \sin x$, it is clear that $y = \sin x$ lies below $y = x$ for $x \geq 0$. So the area between these two curves from $x = 0$ to $x = \pi$ is given by

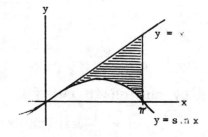

$$[\text{Area}] = \int_0^\pi (x - \sin x)\,dx$$

$$= \left[\frac{x^2}{2} + \cos x\right]\Big|_0^\pi = \left[\frac{\pi^2}{2} + \cos\pi\right] - \left[0 + \cos 0\right]$$

$$= \frac{\pi^2}{2} + (-1) - [0 + 1] = \frac{\pi^2}{2} - 2.$$

CHAPTER 9

TECHNIQUES OF INTEGRATION

9.1 Integration by Substitution

The technique of integration by substitution is best learned by trial and error. The more problems you work the fewer errors you will make in your first guess at a substitution. The basic strategy is to arrange the integrand in the form $f(g(x))g'(x)$. (This is possible for every problem in this section.) Look for the most complicated part of the integrand that may be expressed as a composite function, and let u be the "inside" function $g(x)$. Then proceed as discussed in the text.

Remember that you can check your answers to an antidifferentiation problem by differentiating your answer. (You ought to do this at least mentally for every indefinite integral you find.) We urge you to work *all* the problems (odd and even) in Section 9.1. Use the solutions in this guide to check your work, *not* to show you how to start a problem.

1.   Let $u = x^2 + 4$. Then $du = \frac{d}{dx}(x^2+4)\,dx = 2x\,dx$. So

$$\int 2x(x^2+4)^5\,dx = \int u^5\,du$$

$$= \frac{1}{6}u^6 + c$$

$$= \frac{1}{6}(x^2+4)^6 + c.$$

7.   Let $u = 4 - x^2$. Then $du = \frac{d}{dx}(4-x^2)\,dx = -2x\,dx$. So,

$$\int x\sqrt{4-x^2}\,dx = \int \left(-\frac{1}{2}\right)(-2)x\sqrt{4-x^2}\,dx$$

$$= -\frac{1}{2} \int \sqrt{4-x^2} \overbrace{(-2x)\ dx}^{du} = -\frac{1}{2} \int \sqrt{u}\ du$$

$$= -\frac{1}{2} \int u^{\frac{1}{2}}\ du = \frac{-\frac{1}{2}\ u^{\frac{3}{2}}}{\frac{3}{2}} + C = -\frac{1}{2}\cdot\frac{2}{3}\ u^{\frac{3}{2}} + C$$

$$= -\frac{1}{3}\ u^{\frac{3}{2}} + C = -\frac{1}{3}(4-x^2)^{\frac{3}{2}} + C.$$

13. Let $u = \ln(2x)$. Then $du = \frac{d}{dx}[\ln(2x)]\ dx = \frac{1}{2x}\cdot 2\ dx = \frac{dx}{x}$. Thus,

$$\int \frac{\ln 2x}{x}\ dx = \int \ln 2x\ \frac{dx}{x} = \int u\ du$$

$$= \frac{u^2}{2} + C = \frac{(\ln 2x)^2}{2} + C.$$

19. $\ln\sqrt{x} = \ln x^{1/2} = \frac{1}{2}\ln x$. This gives us the new form:

$$\int \frac{\ln\sqrt{x}}{x}\ dx = \int \frac{\frac{1}{2}\ln x}{x}\ dx = \frac{1}{2} \int \frac{\ln x}{x}\ dx.$$

So we first need to find $\int \frac{\ln x}{x}\ dx$.

Let $u = \ln x$. Then $du = \frac{d}{dx}\ln x\ dx = \frac{1}{x}\ dx$. Hence,

$$\int \frac{\ln x}{x}\ dx = \int u\ du = \frac{u^2}{2} + C = \frac{(\ln x)^2}{2} + C,$$

and this finally gives us

$$\int \frac{\ln\sqrt{x}}{x}\ dx = \frac{1}{2} \int \frac{\ln x}{x}\ dx = \frac{1}{2}\left[\frac{(\ln x)^2}{2} + C\right] = \frac{(\ln x)^2}{4} + C.$$

25. $\int \frac{1}{x \ln x^2}\ dx = \int \frac{dx}{2x \ln x} = \frac{1}{2} \int \frac{dx}{x \ln x}.$

Let $u = \ln x$. Therefore, $du = \frac{d}{dx}\ln x\ dx = \frac{dx}{x}$, and

$$\frac{1}{2} \int \frac{dx}{x \ln x} = \frac{1}{2} \int \frac{1}{u}\ du = \frac{1}{2}\ln|u| + C = \frac{1}{2}\ln|\ln x| + C.$$

31. If $f'(x) = \dfrac{x}{\sqrt{x^2+9}}$, then $f(x) = \displaystyle\int \dfrac{x}{\sqrt{x^2+9}}\, dx$. Let $u = x^2 + 9$.

Then $du = 2x\, dx$, so

$$\int \dfrac{x}{\sqrt{x^2+9}}\, dx = \int \dfrac{\left(\frac{1}{2}\right)2x}{\sqrt{x^2+9}}\, dx = \dfrac{1}{2}\int \dfrac{du}{\sqrt{u}} = \dfrac{1}{2}\int u^{-\frac{1}{2}}\, du$$

$$= \dfrac{1}{2}\dfrac{u^{\frac{1}{2}}}{\left(\frac{1}{2}\right)} + C = u^{\frac{1}{2}} + C = \sqrt{x^2+9} + C.$$

But $f(4) = 8$, so $8 = \sqrt{4^2+9} + C$ which implies that $C = 3$.
Therefore, $f(x) = \sqrt{x^2+9} + 3$.

37. Let $u = \sin x$. Thus $du = \dfrac{d}{dx}(\sin x)\, dx = \cos x\, dx$. So

$$\int \sin x \cos x\, dx = \int u\, du$$

$$= \dfrac{u^2}{2} + C = \dfrac{(\sin x)^2}{2} + C.$$

43. Let $u = 1 - \sin 5x$. Then,

$$du = \dfrac{d}{dx}(1 - \sin 5x)\, dx = -5 \cos 5x\, dx.$$

So,

$$\int (\cos 5x)\sqrt{1-\sin 5x}\, dx = \int \sqrt{1-\sin 5x}\left(-\dfrac{1}{5}\right)(-5 \cos 5x)\, dx$$

$$= -\dfrac{1}{5}\int \sqrt{u}\, du = -\dfrac{1}{5}\int u^{1/2}\, du$$

$$= -\dfrac{1}{5}\dfrac{u^{3/2}}{\frac{3}{2}} + C = -\dfrac{2}{15}u^{3/2} + C$$

$$= -\dfrac{2}{15}(1 - \sin 5x)^{3/2} + C.$$

9.2 Integration by Parts

Exercises 1-24 are all solved using integration by parts. This technique, like the method of substitution, is learned by working many problems. There is another method for integration by parts that some students may have seen elsewhere. It uses *u* in place of *f(x)* and *dv* in place of *g(x) dx*. Years ago we tried teaching both methods and we found that the one we now present in the text is easier to learn and use.

Once you have mastered integration by parts and by substitution, you are ready to try Exercises 25-36. These problems are harder because you have to decide which method to use!

a. Always consider the method of substitution first. This will work if the integrand is the product of two functions and one of the functions is the derivative of the "inside" of the other function (except maybe for a constant multiple).

b. If the integrand is the product of two unrelated functions, try integration by parts.

c. Some antiderivatives cannot be found by either of the methods we have discussed. We will never give you such a problem to solve, but you should be careful if you look in another text for more problems to work.

1. Let $f(x) = x$, $\quad g(x) = e^{5x}$.

$\qquad f'(x) = 1$, $\quad G(x) = \frac{1}{5}e^{5x}$.

Then
$$\int xe^{5x}\, dx = \frac{1}{5}xe^{5x} - \int \frac{1}{5}e^{5x}\, dx$$

$$= \frac{1}{5}xe^{5x} - \frac{1}{5}\int e^{5x}\, dx$$

$$= \frac{1}{5}xe^{5x} - \frac{1}{5}\cdot\frac{1}{5}e^{5x} + C$$

$$= \frac{1}{5}xe^{5x} - \frac{1}{25}e^{5x} + C.$$

Helpful Hint: The various steps in your calculation of $\int xe^{5x}$ dx should be connected by equals signs, showing how you move from one step to the next. It requires a little more effort to rewrite the first term $\frac{1}{5}xe^{5x}$ on each line, but the solution is not correct otherwise. A careless student might write

$$\int xe^{5x}\, dx = \frac{1}{5}xe^{5x} - \int \frac{1}{5}e^{5x}\, dx$$

$$\boxed{=} \qquad - \frac{1}{5}\int e^{5x}\, dx$$

$$\boxed{\text{errors}} \rightarrow \boxed{=} \qquad - \frac{1}{5}\cdot\frac{1}{5}e^{5x} + C$$

$$\boxed{=} \qquad - \frac{1}{25}e^{5x} + C.$$

Avoid such "shortcuts" in your *homework* as well as on tests. Develop the habit of writing proper mathematical solutions on your homework. This will help to generate correct patterns of thought when you take an exam.

7. Let $f(x) = x$, $\quad g(x) = \dfrac{1}{\sqrt{x+1}} = (x+1)^{-1/2}$,

$\quad f'(x) = 1$, $\quad G(x) = 2(x+1)^{1/2}$.

Then $\displaystyle\int \dfrac{x}{\sqrt{x+1}}\,dx = 2x(x+1)^{1/2} - \int 2(x+1)^{1/2}\,dx$

$$= 2x(x+1)^{1/2} - 2\int (x+1)^{1/2}\,dx$$

$$= 2x(x+1)^{1/2} - 2\cdot\tfrac{2}{3}(x+1)^{3/2} + C$$

$$= 2x(x+1)^{1/2} - \tfrac{4}{3}(x+1)^{3/2} + C.$$

13. Let $f(x) = x$, $\quad g(x) = \sqrt{x+1} = (x+1)^{1/2}$,

$\quad f'(x) = 1$, $\quad G(x) = \tfrac{2}{3}(x+1)^{3/2}$.

Then $\displaystyle\int x\sqrt{x+1}\,dx = \tfrac{2}{3}x(x+1)^{3/2} - \int \tfrac{2}{3}(x+1)^{3/2}\,dx$

$$= \tfrac{2}{3}x(x+1)^{3/2} - \tfrac{2}{3}\int (x+1)^{3/2}\,dx$$

$$= \tfrac{2}{3}x(x+1)^{3/2} - \tfrac{2}{3}\cdot\tfrac{2}{5}(x+1)^{5/2} + C$$

$$= \tfrac{2}{3}x(x+1)^{3/2} - \tfrac{4}{15}(x+1)^{5/2} + C.$$

19. Let $f(x) = \ln 5x$, $\quad g(x) = x$,

$\quad f'(x) = \dfrac{5}{5x} = \dfrac{1}{x}$, $\quad G(x) = \dfrac{x^2}{2}$.

Then $\displaystyle\int x \ln 5x\,dx = \dfrac{x^2}{2}\ln 5x - \int \dfrac{1}{x}\cdot\dfrac{x^2}{2}\,dx$

$$= \dfrac{x^2}{2}\ln 5x - \dfrac{1}{2}\int x\,dx$$

$$= \dfrac{x^2}{2}\ln 5x - \dfrac{1}{2}\cdot\dfrac{1}{2}x^2 + C$$

$$= \dfrac{x^2}{2}\ln 5x - \dfrac{1}{4}x^2 + C.$$

25. Since $x(x+5)^4$ is the product of two functions, neither of which is a multiple of the derivative of the other, we use the technique of integration by parts.

Set $f(x) = x$, $\quad g(x) = (x + 5)^4$,

$\qquad f'(x) = 1$, $\quad G(x) = \frac{1}{5}(x + 5)^5$.

Then $\displaystyle\int x(x + 5)^4\, dx = \frac{1}{5}x(x + 5)^5 - \int \frac{1}{5}(x + 5)^5\, dx$

$$= \frac{1}{5}x(x + 5)^5 - \frac{1}{5}\cdot\frac{1}{6}(x + 5)^6 + C$$

$$= \frac{1}{5}x(x + 5)^5 - \frac{1}{30}(x + 5)^6 + C.$$

31. Since $x^2 + 1$ has $2x$ as a derivative, and $2x$ is a multiple of x, we use substitution to evaluate the integral. Let $u = x^2 + 1$. Then $du = 2x\,dx$. So,

$$\int x\sec^2(x^2 + 1)\, dx = \frac{1}{2}\int 2x\sec^2(x^2 + 1)\, dx = \frac{1}{2}\int \sec^2 u\, du$$

$$= \frac{1}{2}\tan u + C = \frac{1}{2}\tan(x^2 + 1) + C.$$

37. Slope is $\dfrac{x}{\sqrt{x+9}}$, so

$$f'(x) = \frac{x}{\sqrt{x+9}},$$

$$F(x) = \int \frac{x}{\sqrt{x+9}}\, dx.$$

Let $u = x + 9$, so $x = u - 9$, and $du = \frac{d}{dx}(x + 9)\,dx = dx$. So,

$$\int \frac{x}{\sqrt{x+9}}\, dx = \int \frac{u-9}{\sqrt{u}}\, du = \int \left(u^{1/2} - 9u^{-1/2}\right) du$$

$$= \frac{u^{3/2}}{3/2} - \frac{9u^{1/2}}{1/2} + C$$

$$= \frac{2}{3} u^{3/2} - 18u^{1/2} + C$$

$$= \frac{2}{3}(x + 9)^{3/2} - 18(x + 9)^{1/2} + C.$$

$(0,2)$ is on the graph, so at $x = 0$: $f(x) = 2$. Therefore,

$$2 = \frac{2}{3}(0 + 9)^{3/2} - 18(0 + 9)^{1/2} + C,$$

$$2 = \frac{2}{3}(9)^{3/2} - 18(9)^{1/2} + C = 18 - 3(18) + C,$$

$$2 = -36 + C, \text{ or } C = 38.$$

$$\text{Thus } f(x) = \frac{3}{2}(x+9)^{3/2} - 18(x+9)^{1/2} + 38$$

$$= \sqrt{x+9}\left[\frac{2}{3}x + 6 - 18\right] + 38$$

$$= \sqrt{x+9}\left[2x - \frac{4}{3}x + 12\right] + 38$$

$$= \sqrt{x+9}\left[2x - \frac{4}{3}\{x + 9\}\right] + 38$$

$$= 2x\sqrt{x+9} - \frac{4}{3}(x+9)^{3/2} + 38.$$

9.3 Definite Integrals

One purpose of this section is to give you lots of practice choosing between integration by parts and by substitution. Another purpose is to show how these procedures may be simplified somewhat when working with definite integrals.

A common mistake in this section is to make a change of variable in a definite integral and *not* change the limits of integration. For example, here is an *incorrect* evaluation of the integral $\int_0^1 2x(x^2+1)^5\,dx$. Let $u = x^2 + 1$ and $du = 2x\,dx$. Then

$$\int_0^1 (x^2+1)^5 \cdot 2x\,dx \;\boxed{=}\; \int_0^1 u^5\,du \;=\; \frac{1}{6}u^6\Big|_0^1$$

$$\boxed{\text{error}}$$

$$= \frac{1}{6}(1)^6 - \frac{1}{6}(0)^6 = \frac{1}{6}.$$

The error in this calculation was made on the first line. The integral on the left equals 21/2, not 1/6. See Example 1.

Here is another incorrect solution whose final answer is correct. However the work contains two errors and therefore does not justify the final answer. As before, let $u = x^2 + 1$, $du = 2x\,dx$. Then

$$\int_0^1 (x^2+1)^5 \cdot 2x\,dx \;\boxed{=}\; \int_0^1 u^5\,du \;=\; \frac{1}{6}u^6\Big|_0^1$$

$$\boxed{\text{errors}}$$

$$\boxed{=}\; \frac{1}{6}(x^2+1)^6\Big|_0^1 \;=\; \frac{1}{6}(2)^6 - \frac{1}{6}(1)^6$$

$$= \frac{63}{6} = \frac{21}{2}.$$

The symbol $\left. \frac{1}{6}u^6 \right|_0^1$ represents the net change in $\frac{1}{6}u^6$ over the interval $0 \le u \le 1$. This number is not the same as the net change in $\frac{1}{6}(x^2 + 1)^6$ over the interval $0 \le x \le 1$.

The second incorrect solution above may be corrected by omitting the limits of integration at first, that is, by considering an antiderivative instead of a definite integral. Thus

$$\int (x^2 + 1)^5 \cdot 2x\, dx = \int u^5\, du = \frac{1}{6}u^6 + C = \frac{1}{6}(x^2 + 1)^6 + C$$

and hence

$$\int_0^1 (x^2 + 1)^5 2x\, dx = \left. \frac{1}{6}(x^2 + 1)^6 \right|_0^1 = \frac{1}{6}(2)^6 - \frac{1}{6}(1)^6 = \frac{21}{2}.$$

This is the first solution method discussed in Example 1.

1. Let $u = 2x - 3$. Then $du = \frac{d}{dx}(2x - 3)\, dx = 2\, dx$.

   If $x = \frac{3}{2}$, then $u = 2(\frac{3}{2}) - 3 = 0$.

   If $x = 2$, then $u = 2(2) - 3 = 1$.

   Therefore $\displaystyle\int_{3/2}^2 2(2x+3)^{17}\, dx = \int_0^1 u^{17}\, du = \left. \frac{1}{18}u^{18} \right|_0^1$

   $$= \frac{1}{18}(1)^{18} - 0 = \frac{1}{18}.$$

7. Since $(x^2 - 9)$ has $2x$ as a derivative, and since this is a multiple of x, we use the substitution $u = x^2 - 9$. Then $du = 2x\, dx$. If $x = 3$, then $u = 3^2 - 9 = 0$; if $x = 5$, then $u = 5^2 - 9 = 16$.

   $$\int_3^5 x\sqrt{x^2 - 9}\, dx = \frac{1}{2}\int_3^5 2x\sqrt{x^2 - 9}\, dx = \frac{1}{2}\int_0^{16} \sqrt{u}\, du$$

   $$= \left. \frac{1}{2} \cdot \frac{2}{3}u^{3/2} \right|_0^{16} = \frac{1}{3} \cdot 16^{3/2} - 0 = \frac{64}{3}.$$

13. Since x^3 has $3x^2$ as a derivative, and since this is a multiple of x^2, we use the substitution $u = x^3$. Then $du = 3x^2 \, dx$. If $x = 1$, then $u = 1^3 = 1$, and if $x = 3$, then $u = 3^3 = 27$.

$$\int_1^3 x^2 e^{x^3} \, dx = \frac{1}{3} \int_1^3 3x^2 e^{x^3} \, dx = \frac{1}{3} \int_1^{27} e^u \, du$$

$$= \frac{1}{3} e^u \Big|_1^{27} = \frac{1}{3} e^{27} - \frac{1}{3} e = \frac{1}{3}(e^{27} - e).$$

19. Since the integrand is the product of two unrelated functions, we use the technique of integration by parts. Let $f(x) = x$, and $g(x) = \sin \pi x$

$$f'(x) = 1, \quad G(x) = -\frac{1}{\pi} \cos \pi x.$$

Then $\displaystyle \int_0^1 x \sin \pi x \, dx = -\frac{1}{\pi} x \cos \pi x \Big|_0^1 - \int_0^1 -\frac{1}{\pi} \cos \pi x \, dx$

$$= \left[-\frac{1}{\pi} \cos \pi - 0 \right] + \frac{1}{\pi} \int_0^1 \cos \pi x \, dx$$

$$= \frac{1}{\pi} + \frac{1}{\pi} \cdot \frac{1}{\pi} \sin \pi x \Big|_0^1$$

$$= \frac{1}{\pi} + \left[\frac{1}{\pi^2} \sin \pi - \frac{1}{\pi^2} \sin 0 \right]$$

$$= \frac{1}{\pi} + 0 = \frac{1}{\pi}.$$

25. Find where $y = x\sqrt{4-x^2}$ cuts x-axis ($y = 0$), so

$$0 = x\sqrt{4-x^2},$$

$$x = 0, \text{ or } \sqrt{4-x^2} = 0 \Rightarrow 4 - x^2 = 0,$$

$$x^2 = 4, \text{ or } x = \pm 2.$$

Area of portion from $x = -2$ to $x = 0$ is given by

$$\int_{-2}^0 x\sqrt{4-x^2} \, dx.$$

Find $\displaystyle \int x\sqrt{4-x^2} \, dx$, let $u = 4 - x^2$

$$du = \frac{d}{dx}(4 - x^2) \, dx = -2x \, dx.$$

So,

$$\int x\sqrt{4-x^2}\,dx = \int \left(-\frac{1}{2}\right)(-2x)\sqrt{4-x^2}\,dx$$

$$= -\frac{1}{2}\int \sqrt{u}\,du = -\frac{1}{2}\int u^{1/2}\,du$$

$$= -\frac{1}{2}\frac{u^{3/2}}{3/2} = -\frac{1}{3}u^{3/2} + c$$

$$= -\frac{1}{3}\left(4 - x^2\right)^{3/2} + c.$$

Therefore,

$$\int_{-2}^{0} x\sqrt{4-x^2}\,dx = -\frac{1}{3}\left(4 - x^2\right)^{3/2}\Bigg|_{-2}^{0}$$

$$= -\frac{1}{3}(4)^{3/2} + \frac{1}{3}(4 - 4)^{3/2} = -\frac{8}{3}.$$

Area is actually $\frac{8}{3}$ (since area is a positive quantity).
By symmetry Area from $x = 0$ to $x = 2$ is also $\frac{8}{3}$.

$$\text{Total Area is } \frac{16}{3}.$$

9.4 Approximation of Definite Integrals

It will be helpful for you to review Section 6.2 before you study Section 9.4. When you work the exercises in Section 9.4, do not be concerned if your answers differ slightly (in, say, the fourth or fifth significant figure) from the answers in the text. When numerical calculations are made that involve several steps, any decision (ours or yours) to "round off" intermediate answers can affect the final answer. Check with your instructor about how many significant figures you should retain in your calculations.

1. The interval has length $5 - 3 = 2$. If we divide this interval into 5 subintervals of equal length, then each subinterval will have length $\Delta x = \frac{2}{5} = .4$. Therefore the endpoints of the subintervals are:

$$a_0 = 3,$$
$$a_1 = 3 + .4 = 3.4,$$
$$a_2 = 3.4 + .4 = 3.8,$$
$$a_3 = 3.8 + .4 = 4.2,$$
$$a_4 = 4.2 + .4 = 4.6,$$
$$a_5 = 4.6 + .4 = 5.$$

7. If $n = 2$, then $\Delta x = (4 - 0)/2 = 2$. The first midpoint is $x_1 = a + \Delta x/2 = 0 + 2/2 = 1$. The second midpoint is $x_2 = 1 + \Delta x = 1 + 2 = 3$. Since $f(x) = x^2 + 5$, we have $f(1) = 6$, and $f(3) = 14$. Using the midpoint rule we have:

$$\int_0^4 (x^2 + 5)\, dx \approx [f(1) + f(3)]\cdot 2$$

$$= (6 + 14)2$$

$$= 40.$$

If $n = 4$, then $\Delta x = (4 - 0)/4 = 1$. The first midpoint is $x_1 = a + \Delta x/2 = 0 + 1/2 = .5$. Since the other midpoints are spaced 1 unit apart, we have $x_2 = 1.5$, $x_3 = 2.5$, and $x_4 = 3.5$. Hence

$$\int_0^4 (x^2 + 5)\, dx \approx [f(x_1) + f(x_2) + f(x_3) + f(x_4)]\,\Delta x$$

$$= \left\{ [(.5)^2 + 5] + [(1.5)^2 + 5] + [(2.5)^2 + 5] + [(3.5)^2 + 5] \right\}(1)$$

On an exam, some instructors may permit you to stop at this point, because the expression above shows that you know how to use the midpoint rule. (Check with your instructor.) Only arithmetic remains. For homework, of course, you should complete the calculation, and obtain

$$\int_0^4 (x^2 + 5)\, dx \approx \{5.25 + 7.25 + 11.25 + 17.25\}(1)$$

$$= 41.$$

Evaluating the integral directly, we have

$$\int_0^4 (x^2 + 5)\, dx = \left(\tfrac{1}{3}x^3 + 5x \right)\Big|_0^4 = \frac{64}{3} + 20 = 41\tfrac{1}{3}.$$

13. For $n = 3$, we have $\Delta x = (5 - 1)/3 = 4/3$. The endpoints of the three subintervals begin at $a_0 = 1$ and are spaced $\tfrac{4}{3}$ units apart. Thus $a_1 = 1 + \tfrac{4}{3} = \tfrac{7}{3}$, $a_2 = \tfrac{7}{3} + \tfrac{4}{3} = \tfrac{11}{3}$, and $a_3 = \tfrac{11}{3} + \tfrac{4}{3} = \tfrac{15}{3} = 5$. By the trapezoidal rule,

$$\int_1^5 \frac{1}{x^2}\, dx \approx \left[f(1) + 2f\left(\tfrac{7}{3}\right) + 2f\left(\tfrac{11}{3}\right) + f(5) \right]\left(\tfrac{4}{3}\right)\cdot\tfrac{1}{2}$$

$$= \left[1 + 2\cdot\frac{1}{\left(\frac{7}{3}\right)^2} + 2\cdot\frac{1}{\left(\frac{11}{3}\right)^2} + \frac{1}{5^2}\right]\left(\frac{4}{3}\right)\cdot\frac{1}{2}.$$

At this point, all substitutions into the trapezoidal rule are finished; only arithmetic remains. (Check to see if you can stop here on an exam.) The fractions above become quite messy, and it is best to use decimal approximations. We'll use five decimal places. (Check to see how many you should use on an exam.) We have

$$\int_1^5 \frac{1}{x^2}\,dx \approx [1 + .36735 + .14876 + .04](.66667)$$
$$= 1.03741.$$

Evaluating the integral directly produces

$$\int_1^5 \frac{1}{x^2}\,dx = \int_1^5 x^{-2}\,dx = -x^{-1}\Big|_1^5 = -\frac{1}{5} - (-1) = \frac{4}{5} = .8.$$

19. When n = 5, we recommend that you use decimals. In this exercise,

$$\Delta x = (5 - 2)/5 = 3/5 = .6.$$

It is often helpful to draw a picture. This is particularly true here because we need both the endpoints and midpoints of the subintervals, and the numbers involved are not as simple as in some exercises. Draw a line segment and, beginning at one end, mark off five equal subintervals. (This is easier than trying to divide one large interval into five equal parts.) Label the left endpoint $a_0 = 2$, and repeatedly add $\Delta x = .6$ to get the other endpoints.

$$\begin{array}{cccccc} 2 & 2.6 & 3.2 & 3.8 & 4.4 & 5 \end{array}$$

The first midpoint is $2 + \Delta x/2 = 2 + .3 = 2.3$. Repeatedly add $\Delta x = .6$ to get the other midpoints.

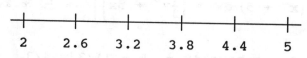

$$\begin{array}{cccccc} 2 & 2.6 & 3.2 & 3.8 & 4.4 & 5 \end{array}$$

Using the midpoint rule for $f(x) = xe^x$, we obtain

$$\int_2^5 xe^x\, dx$$

$$\approx \left[2.3e^{2.3} + 2.9e^{2.9} + 3.5e^{3.5} + 4.1e^{4.1} + 4.7e^{4.7}\right](.6)$$

$$\approx (955.69661)(.6) = 573.41797 = M.$$

For the trapezoidal rule, we have

$$\int_2^5 xe^x\, dx \approx \left[2e^2 + (2)2.6e^{2.6} + (2)3.2e^{3.2} + \right.$$

$$\left. + (2)3.8e^{3.8} + (2)4.4e^{4.4} + 5e^5\right](.6)\cdot\frac{1}{2}$$

$$\approx (2040.36019)(.6)(.5) = 612.10806 = T.$$

We are now prepared to use Simpson's rule.

$$\int_2^5 xe^x\, dx \approx \frac{2M + T}{3} = \frac{2(573.41797) + 612.10806}{3}$$

$$= 586.31467 = S.$$

To evaluate the integral directly, we use the technique of integration by parts. Set

$$f(x) = x, \quad g(x) = e^x,$$

$$f'(x) = 1, \quad G(x) = e^x.$$

Then

$$\int_2^5 xe^x\, dx = xe^x\Big|_2^5 - \int_2^5 e^x\, dx$$

$$= (5e^5 - 2e^2) - e^x\Big|_2^5$$

$$= (742.06580 - 14.77811) -$$

$$(148.41316 - 7.38906)$$

$$= 727.28769 - 141.02410$$

$$= 586.26359.$$

25. We let f(x) represent the distance (in feet) from the shore to the property line as x runs from the top to the bottom of the diagram, and is measured in feet.

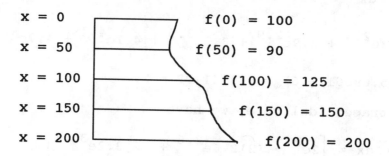

x = 0	f(0) = 100
x = 50	f(50) = 90
x = 100	f(100) = 125
x = 150	f(150) = 150
x = 200	f(200) = 200

So by the trapezoidal rule we have

$$[\text{Area of Property}] = \int_0^{200} f(x)\,dx$$

$$\approx \left[f(0) + 2f(50) + 2f(100) + 2f(150) + f(200) \right] \left(\frac{50}{2} \right)$$

$$= [100 + 180 + 250 + 300 + 200]25$$

$$= (1030)25$$

$$= 25,750 \text{ square feet.}$$

31. a. Note from Figure 8(a) in the text that the height of the triangle is $k - h$. The area of the triangle is $\frac{1}{2}(k - h)\ell$, and the area of the rectangle is $h\ell$. Therefore, the area of the trapezoid is given by

$$A = \frac{1}{2}(k - h)\ell + h\ell$$

$$= \left[\frac{1}{2}k - \frac{1}{2}h + h \right]\ell$$

$$= \left[\frac{1}{2}h + \frac{1}{2}k \right]\ell$$

$$= \frac{1}{2}(h + k)\ell.$$

b. We have $h = f(a_0)$, $k = f(a_1)$, and $\ell = \Delta x$. Therefore,

$$A = \frac{1}{2}[f(a_0) + f(a_1)]\Delta x.$$

c. We know the area of the first trapezoid is:

$$A = \frac{1}{2}[f(a_0) + f(a_1)]\Delta x = [f(a_0) + f(a_1)]\frac{\Delta x}{2}.$$

Similarly, the respective areas of the second, third, and fourth rectangles are

$$A_2 = [f(a_1) + f(a_2)]\frac{\Delta x}{2},$$

$$A_3 = [f(a_2) + f(a_3)]\frac{\Delta x}{2},$$

$$A_4 = [f(a_3) + f(a_4)]\frac{\Delta x}{2}.$$

So the sum of the areas is given by

$$A_1 + A_2 + A_3 + A_4 = \{[f(a_0) + f(a_1)] + [f(a_1) + f(a_2)]$$

$$+ [f(a_2) + f(a_3)] + [f(a_3) + f(a_4)]\}\frac{\Delta x}{2}$$

$$= \{f(a_0) + 2f(a_1) + 2f(a_2) + 2f(a_3) + f(a_4)\}\frac{\Delta x}{2}.$$

This quantity is the value of the trapozoidal rule with $n = 4$ for $\int_a^b f(x)\,dx$. The combined area of the trapezoids is approximately equal to the area under the graph of $f(x)$ from $x = a$ to $x = b$.

9.5   Some Applications of the Integral

The purpose of this section is to show how fairly difficult integration problems can arise easily in applications. The Riemann sum argument in Example 3 is well worth studying, but your instructor may not expect you to reproduce it every time you work an exercise. (Check with your instructor.) Here is the general formula that arises from the analysis in Example 3.

> *A Demographic Model* Suppose that the population density
> of a city t miles from the city center is D(t) thousand
> people per square mile. Then the total number of people
> (in thousands) who live within R miles of the city center
> is approximately equal to
>
> $$\int_0^R 2\pi t D(t)\, dt.$$

1. We use the fact that the present value $= \int_{T_1}^{T_2} K(t)\, e^{-rt} dt.$

    Here $K(t) = 40{,}000$, $T_1 = 0$, $T_2 = 5$, and $r = .08$. There-
    fore,

    $$P.V. = \int_0^5 40000\, e^{-.08t} dt = \left. \frac{40000\, e^{-.08t}}{-.08} \right|_0^5$$

    $$= -500{,}000 \left[e^{-.08(5)} - e^0 \right]$$

    $$= -500{,}000 \left[e^{-.4} - 1 \right] = -500{,}000\, [.67032 - 1]$$

    $$= \$164{,}840.$$

7. We use the formula given in the notes above, with $D(t) = 120e^{-.65t}$, and $R = 5$. The number of people (in thousands) who lived within 5 miles of the city center in 1900 was

    $$\int_0^5 2\pi t \cdot 120 e^{-.65t}\, dt = 240\pi \int_0^5 t e^{-.65t}\, dt$$

    $$= \left. \frac{240\pi t e^{-.65t}}{-.65} \right|_0^5 - 240\pi \int_0^5 \frac{e^{-.65t}}{-.65}\, dt \quad \text{(by integration by parts)}$$

    $$= \frac{-240(5)\pi e^{-3.25}}{.65} - \left. \frac{240\pi\, e^{-.65t}}{(-.65)(-.65)} \right|_0^5$$

    $$= \frac{-1200\pi e^{-3.25}}{.65} - \frac{240\pi}{(.65)(.65)} \left[e^{-3.25} - 1 \right]$$

$$= -5799.863 \ e^{-3.25} - 1784.573 \left[e^{-3.25} - 1 \right]$$

$$\left(e^{-3.25} = .0387742 \right)$$

$$= -224.8850 + 1715.378$$

$$= 1490.493 \text{ thousand people}$$

$$= 1,490,493 \text{ people.}$$

13. a. Area of ring is $2\pi t(\Delta t)$.

   Population density of $40e^{-.5t}$ thousand per square mile.

   Thus the population living in the ring is:

   $$2\pi t(\Delta t) \ 40e^{-.5t} = 80\pi t(\Delta t) \ e^{-.5t} \text{ thousand.}$$

   b. $\dfrac{dP}{dt}$, or $P'(t)$.

   c. It represents the number of people who live between $(5 + \Delta t)$ miles from the city center and 5 miles from the city center.

   d. $P(t+\Delta t) - P(t) = 80\pi t(\Delta t) \ e^{-.5t}$ from (a), so

   $$\frac{P(t+\Delta t) - P(t)}{\Delta t} \approx P't = 80\pi t e^{-.5t}.$$

   e. $\displaystyle\int_{a}^{b} P'(t) \ dt = P(b) - P(a)$, (by the Fundamental Theorem of Calculus)

   $$= \int_{a}^{b} 80\pi t e^{-.5t} dt.$$

9.6 Improper Integrals

Although some instructors may disagree, we feel that the evaluation of an improper integral such as $\displaystyle\int_{0}^{\infty} e^{-x} \ dx$ should *not* use notation such as $-e^{-x}\Big|_{0}^{\infty}$ or $-e^{-\infty} + e^{-0}$. In the context of Section 9.6, the value of a function "at infinity" is not well-defined. The proper notation should involve a limit, such as $\lim\limits_{b \to \infty} (1 - e^{-b})$. You should use two steps to evaluate an improper integral:

i. Compute the net change in the antiderivative over a finite interval, such as from $x = 1$ to $x = b$.

ii. Find the limit of the result in (i) as $b \to \infty$.

1. As b gets large, $\frac{5}{b}$ approaches zero. That is, $\lim_{b \to \infty} \frac{5}{b} = 0$.

7. As b gets large, $\sqrt{b+1}$ gets large, without bound. Hence $\frac{1}{\sqrt{b+1}} = (b+1)^{-1/2}$ approaches zero, and thus $2 - (b+1)^{-1/2}$ approaches 2.

13. As b gets large, e^{3b} gets large, without bound. Hence $\frac{1}{e^{3b}} = e^{-3b}$ approaches zero. That is $\lim_{b \to \infty} e^{-3b} = 0$. Therefore, $\lim_{b \to \infty} \left[2(1 - e^{-3b}) \right] = 2 \left(1 - \lim_{b \to \infty} e^{-3b} \right) = 2(1-0) = 2$.

19. We must calculate the improper integral

$$\int_3^\infty (x+1)^{-3/2} \, dx.$$

We take $b > 3$ and compute

$$\int_3^b (x+1)^{-3/2} \, dx = -2(x+1)^{-1/2} \Big|_3^b = -2(b+1)^{-1/2} + 1.$$

We now consider the limit as $b \to \infty$, noting that $\frac{1}{\sqrt{b+1}}$ approaches zero. Thus

$$\int_3^\infty (x+1)^{-3/2} \, dx = \lim_{b \to \infty} \left(1 - 2(b+1)^{-1/2} \right) = 1 - 2(0) = 1.$$

25. We take $b > 0$ and compute

$$\int_0^b e^{2x} \, dx = \tfrac{1}{2} e^{2x} \Big|_0^b = \tfrac{1}{2} e^{2b} - \tfrac{1}{2}.$$

Now consider the limit as $b \to \infty$. As $b \to \infty$, the number $\tfrac{1}{2} e^{2b} - \tfrac{1}{2}$ can be made larger than any specified number. Therefore, $\int_0^b e^{2x} \, dx$ has no limit as $b \to \infty$, so $\int_0^\infty e^{2x} \, dx$ is divergent.

31. We take b > 0 and compute

$$\int_0^b .01e^{-.01x}\, dx = -e^{-.01x}\Big|_0^b = -e^{-.01b} + 1.$$

As b → ∞, the number $e^{-.01b}$ approaches zero, so $-e^{-.01b}$ approaches zero as well. Therefore,

$$\int_0^\infty .01e^{-.01x}\, dx = \lim_{b\to\infty}\left(-e^{-.01b} + 1\right) = 0 + 1 = 1.$$

37. First consider the indefinite integral $\int xe^{-x^2}\, dx$. Let u = $-x^2$ and du = $-2x\, dx$. Then

$$\int xe^{-x^2}\, dx = -\frac{1}{2}\int e^{-x^2}\cdot(-2x)\, dx = -\frac{1}{2}\int e^u\, du$$

$$= -\frac{1}{2}e^u + C = -\frac{1}{2}e^{-x^2} + C.$$

Now that we have an antiderivative of xe^{-x^2}, we compute

$$\int_0^b xe^{-x^2}\, dx = -\frac{1}{2}e^{-x^2}\Big|_0^b = -\frac{1}{2}e^{-b^2} - \left(-\frac{1}{2}e^0\right)$$

$$= \frac{1}{2} - \frac{1}{2}e^{-b^2}.$$

As b → ∞, $e^{-b^2} = 1/e^{b^2}$ approaches zero. Hence

$$\int_0^\infty xe^{-x^2}\, dx = \lim_{b\to\infty}\left(\frac{1}{2} - \frac{1}{2}e^{-b^2}\right) = \frac{1}{2}.$$

43. We take a < 0 and compute

$$\int_a^0 \frac{6}{(1-3x)^2}\, dx = 6\int_a^0 (1-3x)^{-2}\, dx = 6\cdot\frac{1}{3}(1-3x)^{-1}\Big|_a^0$$

$$= 2 - \frac{2}{1-3a}.$$

Now consider the limit as a → −∞. As a → −∞ the number 1 − 3a gets large, without bound, so $\frac{2}{1-3a}$ approaches zero. So

$$\int_{-\infty}^{0} \frac{6}{(1-3x)^2}\, dx = \lim_{a \to -\infty} \left(2 - \frac{2}{1-3a}\right) = 2.$$

49. We must evaluate the improper integral $\int_{0}^{\infty} 5000e^{-.1t}\, dt$.

Take $b > 0$ and compute

$$\int_{0}^{b} 5000e^{-.1t}\, dt = -50{,}000e^{-.1t}\Big|_{0}^{b} = -50{,}000(e^{-.1b} - 1)$$

$$= 50{,}000 - 50{,}000e^{-.1b}.$$

As $b \to \infty$, $e^{-.1b}$ approaches zero and $50{,}000 - 50{,}000e^{-.1b}$ approaches $50{,}000$. Therefore, $\int_{0}^{\infty} 5000e^{-.1t}\, dt = 50{,}000$.

The capital value generated is $50,000.

Chapter 9: Supplementary Exercises

Sections 9.1-9.3 and 9.6 provide the tools for solving many of the problems in the next three chapters. It is essential that you master the techniques in Sections 9.1 and 9.2. Supplementary Exercises 19-36 will help you learn to recognize which technique to use.

1. Since $3x^2$ has $6x$ as a derivative, and this is a multiple of x, we use the technique of substitution. Let $u = 3x^2$; then $du = 6x\, dx$. We then have

$$\int x \sin 3x^2\, dx = \frac{1}{6}\int 6x \sin 3x^2\, dx$$

$$= \frac{1}{6}\int \sin u\, du$$

$$= -\frac{1}{6}\cos u + C$$

$$= -\frac{1}{6}\cos 3x^2 + C.$$

7. The derivative of $4 - x^2$ is a multiple of x (the factor next to the square root. So make the substitution

$$u = 4 - x^2, \quad du = -2x\, dx.$$

Then

$$\int x\sqrt{4-x^2}\,dx = -\frac{1}{2}\int (-2x)\sqrt{4-x^2}\,dx$$

$$= -\frac{1}{2}\int \sqrt{u}\,du$$

$$= -\frac{1}{2}\cdot\frac{2}{3}u^{3/2} + C$$

$$= -\frac{1}{3}(4 - x^2)^{3/2} + C.$$

13. First simplify the integrand, using the algebraic property $\ln x^2 = 2\ln x$:

$$\int \ln x^2\,dx = 2\int \ln x\,dx$$

Then recall that $\int \ln x\,dx$ is easily found by integration by parts. Use $f(x) = \ln x$ and $g(x) = 1$. The details are in Example 7 of Section 9.2. Using the results of that example, we have

$$\int \ln x^2\,dx = 2\int \ln x\,dx = 2(x\ln x - x) + C.$$

Helpful Hint: Integrals of the form $\int x^k \ln x\,dx$ are found by integration by parts, with $f(x) = \ln x$ and $g(x) = x^k$.

19. Since the integrand is the product of two functions, x and e^{2x}, neither of which is a multiple of the derivative of the other, we use integration by parts. Set

$$f(x) = x, \qquad g(x) = e^{2x}.$$

25. Since the integrand is the product of two functions, e^{-x} and $(3x - 1)^2$, and neither is a multiple of the derivative of the other, we use integration by parts. Set

$$f(x) = (3x - 1)^2, \qquad g(x) = e^{-x}.$$

In order to complete the problem, we must integrate by parts a second time. Note that if we interchange $f(x)$ and $g(x)$ above, integration by parts will not yield a solution.

31. We observe that the derivative of $x^2 + 6x$ is $2x + 6$, which is a multiple of $x + 3$. Thus we use the technique of substitution, setting $u = x^2 + 6x$.

37. Since $x^2 + 1$ has $2x$ as a derivative, we use the technique of substitution. Let $u = x^2 + 1$. Then $du = 2x\,dx$. If $x = 0$, then $u = 1$, and if $x = 1$, then $u = 2$. So we have

$$\int_0^1 \frac{2x}{(x^2+1)^3}\,dx = \int_1^2 \frac{1}{u^3}\,du = \int_1^2 u^{-3}\,du$$

$$= -\frac{1}{2}u^{-2}\Big|_1^2 = -\frac{1}{2}\left(\frac{1}{4}\right) + \frac{1}{2} = \frac{3}{8}.$$

43. The interval has length $9 - 1 = 8$, so with $n = 4$ we have $\Delta x = \frac{8}{4} = 2$. The subintervals have endpoints $a_0 = 1$, $a_1 = 3$, $a_2 = 5$, $a_3 = 7$, and $a_4 = 9$, and midpoints $x_1 = 2$, $x_2 = 4$, $x_3 = 6$, and $x_4 = 8$.

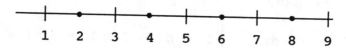

Using the midpoint rule we have, with $f(x) = \dfrac{1}{\sqrt{x}} = x^{-1/2}$,

$$\int_1^9 f(x)\,dx \approx M = [f(2) + f(4) + f(6) + f(8)]\cdot 2$$

$$= [.70711 + .50000 + .40825 + .35355]\cdot 2$$

$$= 3.93782.$$

Using the trapezoidal rule, with $f(x) = \dfrac{1}{\sqrt{x}} = x^{-1/2}$, we have

$$\int_2^9 f(x)\,dx \approx T = [f(1) + 2f(3) + 2f(5) + 2f(7) + f(9)]\cdot\frac{2}{2}$$

$$= [1 + 1.15470 + .89443 + .75593 + .33333]$$

$$= 4.13839.$$

Using Simpson's rule, we have

$$\int_1^9 \frac{1}{\sqrt{x}} \, dx \approx \frac{2M + T}{3} = 4.00468.$$

49. Since the derivative of $x^2 + 4x - 2$ is $2x + 4$, which is a multiple of $x + 2$, we use the technique of substitution. Let $u = x^2 + 4x - 2$; then $du = 2x + 4 \, dx$.

$$\int \frac{x+2}{x^2+4x-2} \, dx = \frac{1}{2} \int \frac{2x+4}{x^2+4x-2} \, dx = \frac{1}{2} \int \frac{1}{u} \, du$$

$$= \frac{1}{2} \ln|u| = \frac{1}{2} \ln|x^2+4x-2| + C.$$

Hence

$$\int_1^b \frac{x^2+2}{x^2+4x-2} \, dx = \frac{1}{2} \ln|x^2+4x-2| \ \Big|_1^b$$

$$= \frac{1}{2} \ln|b^2+4b-2| - \frac{1}{2} \ln 3.$$

As $b \to \infty$, the number $|b^2+4b-2|$ gets large, without bound, and so $\ln|b^2+4b-2|$ gets large, without bound. Therefore $\frac{1}{2} \ln|b^2+4b-2| - \frac{1}{2} \ln 3$ can be made larger than any specific number, $\int_1^b \frac{x+2}{x^2+4x-2} \, dx$ has no limit as $b \to \infty$, and $\int_1^\infty \frac{x+2}{x^2+4x-2} \, dx$ is divergent.

55. We have $K(t) = 50e^{-.08t}$, $r = 12\% = .12$, and $T = 4$. We use the formula stating that present value equals

$$\int_0^T K(t) e^{-rt} \, dt = \int_0^4 50e^{-.08t} e^{-.12t} \, dt$$

$$= \int_0^4 50e^{-.2t} \, dt = -250e^{-.2t} \ \Big|_0^4$$

$$\approx -112.332 + 250$$

$$= 137.668.$$

The present value of the continuous stream of income over the next 4 years is $137,668 (to the nearest dollar).

CHAPTER 10

DIFFERENTIAL EQUATIONS

10.1 Solutions of Differential Equations

This section lays the foundation for the chapter. You should read the text and the examples several times. The high point of the chapter is Section 10.5 where you will use differential equations to gain insights into a variety of applications. Our students find this material extremely interesting, and we know that they can master it if they are given adequate preparation. So each section, beginning with 10.1, includes a few problems that teach how to read and understand a verbal problem about a differential equation.

Example 5 and Exercises 13 and 14 are extremely important. Study the example carefully and then try the exercises. You may have trouble with Exercise 11, but your objective here is to simply *try* it. We'll discuss the solution, and then you can try a similar problem in Section 10.2.

1. We replace y by $f(t) = \frac{3}{2}e^{t^2} - \frac{1}{2}$, and y' by $f'(t) = 3te^{t^2}$.

    Then $y' - 2ty = 3te^{t^2} - 2t\left[\frac{3}{2}e^{t^2} - \frac{1}{2}\right]$

    $$= 3te^{t^2} - [3te^{t^2} - t] = t.$$

    This is true for all t, so $f(t) = \frac{3}{2}e^{t^2} - \frac{1}{2}$ is a solution.

7. The derivative of the constant function $f(t) = 3$ is the constant function $f'(t) = 0$. If we replace y by $f(t) = 3$ and y' by $f'(t) = 0$, then

    $$y' = 6 - 2y,$$
    $$0 = 6 - 2(3).$$

    Both sides are zero for all t, so $f(t) = 3$ is a solution.

13. Let f(t) be the amount of capital invested. Then f'(t) is the rate of net investment. We are told that

$$\begin{bmatrix} \text{rate of net} \\ \text{investment} \end{bmatrix} \text{[is proportional to]} \left\{ C - \begin{bmatrix} \text{capital} \\ \text{investment} \end{bmatrix} \right\}.$$

So there is a constant of proportionality, k, such that

$$f'(t) = k\{C - f(t)\}.$$

To determine the sign of k, suppose that at some time f(t) is larger than C. Then C − f(t) is negative. Should f'(t) be positive or negative? That is, should f(t) be increasing or decreasing? Well, if the amount of f(t) is larger than the optimum level C, then it is reasonable to want f(t) to decrease down towards C. So when C − f(t) is negative, we want f'(t) to be negative, too. This means that k must be positive.

$$f'(t) = k\{C - f(t)\},$$
$$\text{[neg]} = \text{[pos]} \cdot \text{[neg]}.$$

You would arrive at the same conclusion that k is positive if you supposed that f(t) is less than C. The differential equation is simplified by writing y for f(t) and y' for f'(t), namely,

$$y' = k(C - y), \quad k > 0.$$

10.2 Separation of Variables

This is probably the most difficult section in Chapter 10 because it requires substantial integration and algebraic skills. Each example illustrates one or two situations you will encounter in the exercises. After you study Examples 1, 2, and 3, you may work on Exercises 1-6, 13-15, and 17, respectively.

Remember to use the examples properly. Study them carefully and then try to work the exercises *without referring to the examples.* If you really get stuck, take a peek at the appropriate example, but do not study the entire solution. As a last resort, peek at a solution in this study guide if a solution is available. You simply *must* train yourself to work independently. If you yield to the temptation to "copy" our solutions, you will not survive an examination on this material.

Exercises 33-38 are here to help you prepare for Section 10.5. You will be rewarded later if you spend some time on these exercises now.

1. Separating the variables, we have

$$y^2 \frac{dy}{dt} = 5 - t,$$

$$\int y^2 \frac{dy}{dt} \, dt = \int (5 - t) \, dt,$$

$$\int y^2 \, dy = \int (5 - t) \, dt.$$

One antiderivative of y^2 is $\frac{1}{3}y^3$. One antiderivative of $5 - t$ is $5t - \frac{1}{2}t^2$. Since antiderivatives of the same function differ by a constant, we have

$$\frac{1}{3}y^3 = 5t - \frac{1}{2}t^2 + C_1,$$

for some constant C_1. Then

$$y^3 = 3(5t - \frac{1}{2}t^2 + C_1).$$

Since C_1 is arbitrary, so is $3C_1$, and it is easier to write C in place of $3C_1$. Therefore

$$y^3 = 15t - \frac{3}{2}t^2 + C,$$

and

$$y = (15t - \frac{3}{2}t^2 + C)^{1/3}.$$

7. $\frac{dy}{dt} = \frac{t^2}{y^2}e^{t^3}$. By separating the variables,

$$y^2 \frac{dy}{dt} = t^2 e^{t^3}$$

$$\int y^2 \frac{dy}{dt} \, dt = \int t^2 e^{t^3} \, dt$$

$$\int y^2 \, dy = \int t^2 e^{t^3} \, dt.$$

One antiderivative of y^2 is $\frac{1}{3}y^3$, and similarly an anti-derivative of $t^2 e^{t^3}$ is $\frac{1}{3}e^{t^3}$. (This is found by a substitution, $u = t^3$.) So

$$\frac{1}{3}y^3 = \frac{1}{3}e^{t^3} + C.$$

Hence,

$$y^3 = e^{t^3} + C \text{ (another constant } C),$$

and finally,

$$y = \left(e^{t^3} + C\right)^{1/3}.$$

13. $\frac{dy}{dt} = 2 - y.$

The constant function y = 2 is a solution because it makes both sides of the equation zero for all t. (The left-hand side is zero because the derivative of a constant function is zero.)

Now assuming y ≠ 2, we may divide by 2-y:

$$\frac{1}{2-y}\frac{dy}{dt} = 1,$$

so,

$$-\frac{1}{y-2}\frac{dy}{dt} = 1, \text{ or } \frac{1}{y-2}\frac{dy}{dt} = -1,$$

$$\int \frac{1}{y-2}\frac{dy}{dt}\,dt = \int -1\,dt,$$

$$\ln|y-2| = -t + C,$$

$$|y-2| = e^{-t+C} = e^C \cdot e^{-t}.$$

In Section 10.4 we shall see that if y ≠ 2, then y - 2 will be either positive for all t or negative for all t. Consequently

$$y - 2 = e^C \cdot e^{-t} \text{ or } y - 2 = -e^C \cdot e^{-t},$$

$$y = 2 + e^C \cdot e^{-t} \text{ or } y = 2 - e^C \cdot e^{-t}.$$

Since e^C is an arbitrary positive constant and $-e^C$ is an arbitrary negative constant, we may write these two types of solutions, together with the constant solution y = 2 + $0 \cdot e^{-t}$, in the form

$$y = 2 + Ae^{-t}, \text{ A any constant.}$$

(The "C" in the answers in the text is the same as this "A".)

19. $y' = 2t\,e^{-2y} - e^{-2y}, \qquad y(0) = 3.$

The equation can be rewritten as

$$\frac{dy}{dt} = 2t\,e^{-2y} - e^{-2y} = e^{-2y}(2t - 1).$$

Thus applying the method of separation of variables,

$$\frac{1}{e^{-2y}}\frac{dy}{dt} = 2t - 1, \qquad \left(\frac{1}{e^{-2y}} = e^{2y}\right),$$

$$\int e^{2y} \frac{dy}{dt}\, dt = \int (2t - 1)\, dt,$$

$$\int e^{2y}\, dy = \int (2t - 1)\, dt,$$

$$\frac{1}{2} e^{2y} = t^2 - t + C.$$

Hence, $e^{2y} = 2t^2 - 2t + C.$ (This C is actually 2 times the old C.) Take the logarithm of both sides:

$$\ln e^{2y} = 2y \ln e = 2y = \ln(2t^2 - 2t + C).$$

Therefore,

$$y = \frac{1}{2} \ln (2t^2 - 2t + C).$$

Since $y(0) = 3$ we can find C: i.e., if $t = 0$, $y = 3$, giving,

$$3 = \frac{1}{2} \ln (2 \cdot 0^2 - 2(0) + C) = \frac{1}{2} \ln C,$$

$$6 = \ln C,$$

$$e^6 = C.$$

So finally,

$$y = \frac{1}{2} \ln(2t^2 - 2t + e^6).$$

25. $\frac{dy}{dt} = \frac{t+1}{ty},$ $t > 0,$ $y(1) = -3.$

Separating the variables:

$$y \frac{dy}{dt} = \frac{t+1}{t} = 1 + \frac{1}{t},$$

$$\int y \frac{dy}{dt}\, dt = \int \left(1 + \frac{1}{t}\right) dt,$$

$$\int y\, dy = \int \left(1 + \frac{1}{t}\right) dt,$$

therefore,

$$\frac{y^2}{2} = t + \ln|t| + C.$$

Since $t > 0$ we can rewrite $\ln|t|$ as $\ln t$, so

$$\frac{y^2}{2} = t + \ln t + C,$$

$$y^2 = 2t + 2\ln t + C$$

(C is now 2 times the value of the old C).

$$y = \pm\sqrt{2t + 2\ln t + C}.$$

Since $y(1) = -3$, and since a square root cannot be negative, we use the "minus" form:

$$y = -\sqrt{2t + 2\ln t + C},$$

i.e., $-3 = -\sqrt{2(1) + 2\ln(1) + C}$,

$$-3 = -\sqrt{2 + 0 + C}, \quad \text{(square both sides)}$$
$$9 = 2 + C, \quad \text{or } C = 7.$$

Therefore,

$$y = -\sqrt{2t + \ln t^2 + 7}.$$

31. Clearly $y = 0$ is a solution since this makes both sides of the equation equal zero. If $y \neq 0$ we may divide by y to obtain

$$\frac{1}{y}\frac{dy}{dp} = -\frac{1}{2} \cdot \frac{1}{p+3},$$

$$\int \frac{1}{y}\frac{dy}{dp}\, dp = \int -\frac{1}{2} \cdot \frac{1}{p+3}\, dp,$$

$$\int \frac{1}{y}\, dy = -\frac{1}{2}\int \frac{1}{p+3}\, dp,$$

$$\ln|y| = -\frac{1}{2}\ln|p+3| + C, \ C \text{ a constant}$$

$$\ln|y| = \ln|p+3|^{-1/2} + C,$$

$$|y| = e^{\ln|p+3|^{-1/2}+C} = |p+3|^{-1/2} \cdot e^C,$$

$$y = \pm e^C |p+3|^{-1/2}.$$

The general solution (including the constant solution) has the form

$$y = A|p+3|^{-1/2}, \ A \text{ any constant.}$$

Note, however, that both the price p and the sales volume y should be positive quantities. So the only solutions that make some sense economically have the form

$$y = A(p+3)^{-1/2}, \ A > 0.$$

37. $\dfrac{dy}{dt} = -ay \ln \dfrac{y}{b}$. Note that y cannot be zero, because the logarithm would not be defined. However, $\ln \dfrac{y}{b}$ is zero when $y = b$. So the only constant solution is $y = b$. If $y \ne b$, we may separate the variables:

$$\frac{1}{y \ln \frac{y}{b}} \cdot \frac{dy}{dt} = -a,$$

$$\int \frac{1}{y \ln \frac{y}{b}} \cdot \frac{dy}{dt}\, dt = \int -a\, dt,$$

$$\int \frac{1}{y \ln \frac{y}{b}} \frac{dy}{dt} = \int -a\, dt = -at + C.$$

To evaluate the left-hand side, let

$$u = \ln \frac{y}{b} \quad \Rightarrow \quad du = \frac{1}{\left(\frac{y}{b}\right)} \cdot \frac{1}{b}\, dy = \frac{1}{y}\, dy.$$

Then

$$\int \frac{1}{\ln \frac{y}{b}} \cdot \frac{1}{y}\, dy = \int \frac{1}{u}\, du = \ln|u| + C,$$

$$= \ln \left| \ln \frac{y}{b} \right| + C.$$

Therefore, setting the two sides equal,

$$\ln \left| \ln \frac{y}{b} \right| = -at + C \quad \text{(different } C\text{)}.$$

Next, take the exponential of both sides to produce:

$$\left| \ln \frac{y}{b} \right| = e^{-at+C} = e^{C} \cdot e^{-at} = Ce^{-at} \quad (C = e^{C}),$$
$$\text{(different } C \text{ again)}$$

$$\left| \ln \frac{y}{b} \right| = Ce^{-at}, \text{ where } C \text{ is positive,}$$

$$\ln \frac{y}{b} = \pm Ce^{-at}.$$

Let's write $\pm C$ as simply C, where now C can be positive or negative. The constant solution $y = b$ corresponds to $C = 0$. So the general solution, with an arbitrary C, is

$$\frac{y}{b} = e^{Ce^{-at}}, \quad \text{and} \quad y = be^{Ce^{-at}}.$$

10.3 Numerical Solution of Differential Equations

Although this section is not essential for later work, it will focus your attention on ideas that will be of great help in Section 10.4. In particular, you should study Exercises 1-4, even if Section 10.3 is omitted. Likewise, Exercises 11(a) and 12(a) should be attempted as part of your ongoing program of preparation for Section 10.5.

1. If $f(t)$ is a solution of $y' = ty - 5$, then

$$f'(t) = tf(t) - 5,$$

for all t in the domain of $f(t)$. The graph of $f(t)$ passing through $(2,4)$ means that $f(2) = 4$. Setting $t = 2$ in the equation above, we find that

$$f'(2) = 2f(2) - 5,$$
$$= 2 \cdot 4 - 5 = 3.$$

So the slope of the graph is 3 at $t = 2$.

Helpful Hint: Make sure you understand the solutions to Exercises 1-4. They make good test questions and they prepare you for both Euler's method and the theory in Section 10.4.

7. Here $g(t,y) = 2t - y + 1$, $a = 0$, $b = 2$, $y(0) = 5$, and $h = (2-0)/4 = \frac{1}{2}$. Starting with $(t_0, y_0) = (0,5)$, we find $g(0,5) = 2(0) - 5 + 1 = -4$. Thus,

$$t_1 = \frac{1}{2}, \qquad y_1 = 5 + (-4) \cdot \frac{1}{2} = 3.$$

Next, $g(\frac{1}{2}, 3) = 2(\frac{1}{2}) - 3 + 1 = -1$, so

$$t_2 = 1, \qquad y_2 = 3 + (-1)\frac{1}{2} = \frac{5}{2}.$$

Next $g(1, \frac{5}{2}) = 2(1) - \frac{5}{2} + 1 = \frac{1}{2}$, so

$$t_3 = \frac{3}{2}, \qquad y_3 = \frac{5}{2} + (\frac{1}{2})\frac{1}{2} = \frac{11}{4}.$$

And finally, $g(\frac{3}{2}, \frac{11}{4}) = 2(\frac{3}{2}) - \frac{11}{4} + 1 = \frac{5}{4}$, so

$$t_4 = 2, \qquad y_4 = \frac{11}{4} + (\frac{5}{4})\frac{1}{2} = \frac{27}{8}.$$

Thus the approximation to the solution $f(t)$ is given by the polygonal path shown in the answer section of the text. The last point $(2, \frac{27}{8})$ is close to the graph of $f(t)$ at $t = 2$, so $f(2) \approx \frac{27}{8}$.

Helpful Hint: The fractions in Exercise 7 were rather simple. In many cases decimals are easier to use, particularly when h is smaller than $\frac{1}{4}$. Of course, a calculator becomes indispensable in such cases.

11. This problem is part of our program to prepare you for Section 10.5. You should definitely try part (a) even if the problem is not assigned. Before you look at the answer in your text, try to determine whether the constant of proportionality is positive or negative.

13. The graph is in the solutions at the end of the text.

10.4 Qualitative Theory of Differential Equations

We hope you get to study this section! Some instructors omit it because they think it is too hard for first-year calculus students. (No other book at this level attempts to teach the material.) But we have found that our students do as well on the qualitative theory as on almost any material in the second half of the text. The secret is to spend about one week on this section.

One of the difficulties here is learning to sketch yz-graphs where the y-axis is horizontal. The difficulty is real though it is mainly psychological. To help you get used to yz-graphs, we have included them for Exercises 1-14. Study them there so you will be able to produce similar graphs in Exercises 15-31.

1. From y' = 2y - 6 we obtain the function g(y) = 2y - 6. The graph of z = g(y) is shown in Fig. 1a. Setting 2y - 6 = 0, we find y = 3. Thus g(y) has a zero when y = 3. Therefore the constant function y = 3 is a solution of the differential equation. See Fig. 1b.

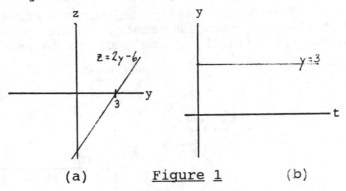

(a)   <u>Figure 1</u>   (b)

To sketch the solution corresponding to y(0) = 1, we locate this initial value on the y-axes in Fig. 2a and Fig. 2b, and we note that the z-coordinate in Fig. 2a is negative when y = 1.   That is, the derivative is negative when y = 1.   So we place a downward arrow at the initial point in Fig. 2b.

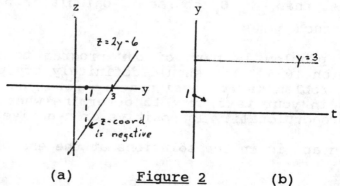

(a)   __Figure 2__   (b)

From Fig. 2b, the y-values will decrease as time passes, so y will move to the __left__ on the yz-graph.   See Fig. 3a. As a result, the z-coordinate on the graph of z = 2y - 6 will become more negative.   That is, the slope of the solution curve will become more negative.   Thus the solution curve is concave down, as in Fig. 3b.

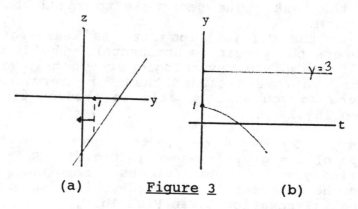

(a)   __Figure 3__   (b)

To obtain the graph of the solution satisfying y(0) = 4, we make the following sketches.   (The verbal description is omitted because it is too tedious to read.)

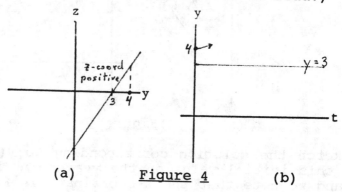

(a)   __Figure 4__   (b)

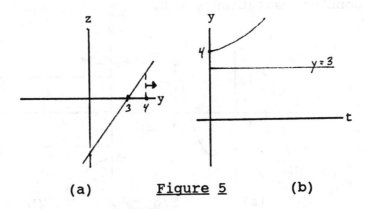

(a)   <u>Figure 5</u>   (b)

The z-coordinate becomes more positive in Fig. 5a, so the solution curve is concave up in Fig. 5b.

7. First we sketch the graph of $z = y^3 - 9y$. We find the zeros of $g(y) = y^3 - 9y$ by setting $g(y) = 0$ and solving for y.

$$y^3 - 9y = 0,$$
$$y(y^2 - 9) = 0,$$
$$y(y + 3)(y - 3) = 0.$$

The graph of $z = y^3 - 9y$ crosses the y-axis at $y = 0$, $y = \pm 3$. Next, to find where the graph has a relative maximum and relative minimum, we set $\frac{dz}{dy} = 0$ and solve for y.

$$\frac{d}{dy}(y^3 - 9y) = 3y^2 - 9 = 0.$$

Thus $3y^2 = 9$, $y^2 = 3$, and $y = \pm\sqrt{3}$. See Fig. 1a. The constant solutions are shown in Fig. 1b.

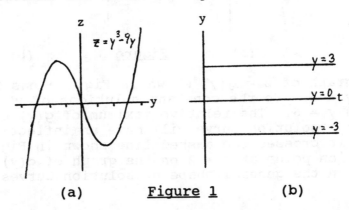

(a)   <u>Figure 1</u>   (b)

Figure 2 shows the sketches needed to obtain the graphs of the solutions such that $y(0) = -4$, $y(0) = -1$, and $y(0) = 4$. Note that the solution where $y(0) = -1$ cannot cross

10-11

the constant solution y = 0.

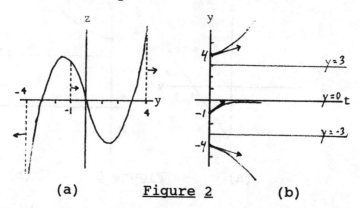

(a)   **Figure 2**   (b)

Figure 3 shows the work for the solution with y(0) = 2. Since the initial z-coordinate in Fig. 3a is negative, the y-values of the solution curve will decrease and y will move to the left on the yz-graph. At first, the z-coordinate will become more negative (i.e., the slope of the solution curve will become more negative). Then, as y moves to the left past $y = \sqrt{3}$, the z-coordinates on the yz-graph will become less negative (i.e., the slope of the solution curve will become less negative). Thus the solution curve will have an inflection point at $y = \sqrt{3}$, as in Fig. 3b.

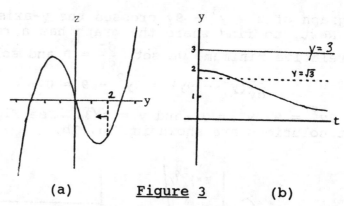

(a)   **Figure 3**   (b)

13. The graph of z = g(y) shown in Fig. 1a has zeros at y = 1 and y = 6. So the constant solutions of y' = g(y) are y = 1 and y = 6. The relative maximum of g(y) occurs when y = 4, so a solution curve will have an inflection point whenever it crosses the dashed line shown in Fig. 1b. The inflection point at y = 2 on the graph of g(y) has no influence on the general shape of solution curves, so we ignore it.

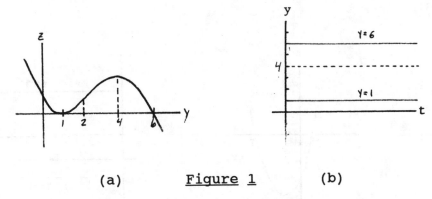

(a)   Figure 1   (b)

Figure 2 shows the work for the solution of $y' = g(y)$ such that $y(0) = 0$, $y(0) = 1.2$, $y(0) = 5$, and $y(0) = 7$.

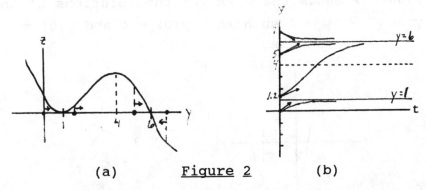

(a)   Figure 2   (b)

Warning: Although the function $g(y)$ in Exercise 13 has a minimum at $y = 1$, the minimum value is zero. Notice that rule 7 on page 631 applies only to <u>nonzero</u> relative maximum or minimum points.

19. The graph of $z = y^2 - 3y - 4$ is obviously a parabola. It opens upward because the coefficient of y^2 is positive. To find where the graph crosses the y-axis, we solve

$$y^2 - 3y - 4 = 0,$$
$$(y+1)(y-4) = 0,$$
$$y = -1, \text{ or } y = 4.$$

Setting $\frac{dz}{dy} = 0$, we find that $2y - 3 = 0$ and $y = \frac{3}{2} = 1.5$. Thus the parabola has a minimum at $y = 1.5$. This is enough information to produce the graph in Fig. 1a and the initial sketch in Fig. 1b.

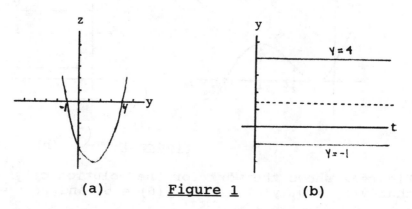

(a)   Figure 1   (b)

Figure 2 shows the work for the solutions of the equation
$y' = y^2 - 3y - 4$ such that $y(0) = 0$ and $y(0) = 3$.

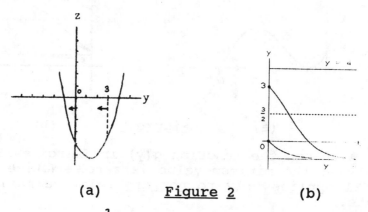

(a)   Figure 2   (b)

25. The graph of $z = \frac{1}{y}$ is shown in Fig. 1a.   The graph does
    not cross the y-axis, so there are no constant solutions
    of $y' = \frac{1}{y}$.   The solution satisfying $y(0) = 1$ is an in-
    creasing curve because the z-coordinate of the initial
    point on the yz-graph is positive. However, as y increas-
    es on the yz-graph, the z-coordinates become less posi-
    tive; i.e., the slopes of the solution curve become less
    positive. A similar solution holds for the solution sat-
    isfying $y(0) = -1$, where the slopes are negative and be-
    come less negative. See Fig. 1b.

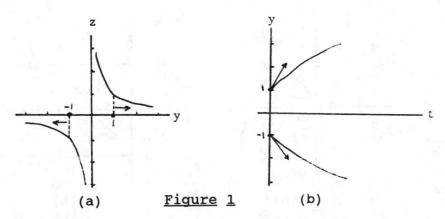

(a)   <u>Figure 1</u>   (b)

Warning: Exercise 27 has a feature similar to Exercise 13.

31. Let $f(t)$ be the height of the sunflower at time t. Then $f'(t)$ is the rate at which the sunflower is growing at time t. We are told that this rate is proportional to the product of its height and the difference between its height at maturity and its current height. If we let H be the sunflower's height at maturity, we have

$$f'(t) = kf(t)[H - f(t)] \qquad (1)$$

or equivalently,

$$y' = ky(H - y). \qquad (2)$$

Before we can sketch a solution, we must determine whether k is positive or negative. Clearly, $f(t)$ is increasing, so $f'(t)$ is positive. Also, $f(t)$ is less than H, so the factor $H - f(t)$ is positive. From this it follows that k must be positive.

From (2) we must sketch,

$$z = ky(H - y) = kHy - ky^2.$$

The graph of this equation is a parabola which opens down, with zeros at 0 and H, and maximum at $H/2$. See Fig. 1(a). The initial height of the sunflower is $y(0)$, which is slightly greater than 0. Since z is positive here, this solution is increasing as it leaves the initial point. As y moves to the right, the z-coordinate becomes more positive until y reaches $H/2$. After than, the z-coordinate becomes less positive, and hence the curve has an inflection point at $H/2$, and is concave down after than point. Also, it is clear from equation (1) that the constant functions $y = 0$ and $y = H$ are solutions. See Fig. 1b.

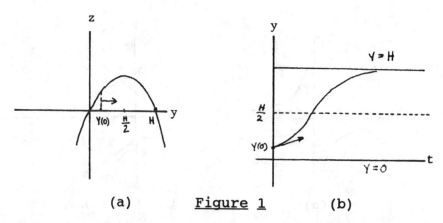

(a)   Figure 1   (b)

Helpful Hint: Exercise 31 is a key exercise. Work on it now! You'll see more exercises like this in Section 10.5.

10.5 Applications of Differential Equations

Here we are! If you've been faithful in your work on the "word problems" in Sections 10.1 to 10.4, you should be ready for Section 10.5. The exercises fall into two categories: (1) problems involving one constant of proportionality, and (2) "one-compartment" problems. The first category is analyzed carefully in the first two pages of the section. Most of Section 5.4 concerned applications of the first category of problems. One of the most important applications there involved logistic growth, and this subject is discussed again in Section 10.5.

One-compartment problems are very common in applications. Multi-compartment problems also arise. You can read about them in *Mathematical Techniques for Physiology and Medicine*, by William Simon (New York: Academic Press, 1972). This book was at one time the text for a course at the Rochester School of Medicine and Dentistry.

The material on population genetics was suggested to us by one of our students. While in our class, she was taking a course on population genetics where she saw a qualitative analysis of certain differential equations. Only the yz-graphs were shown in her textbook, *Population Genetics*, by C. C. Li (Chicago: University of Chicago Press, 1955). Our student used her knowledge from our calculus course to provide the ty-graphs of the solution curves. Now you, too, can have a glimpse into this fascinating topic.

1. Let $y = f(t)$ be the percentage of the population having the information at time t. Then $0 \leq f(t) \leq 100$, and $100 - f(t)$ is the percentage that does not have the information. We are told that

$$\begin{bmatrix} \text{the rate of} \\ \text{spread of} \\ \text{information} \end{bmatrix} \begin{bmatrix} \text{is} \\ \text{proportional} \\ \text{to} \end{bmatrix} \begin{bmatrix} \text{percentage} \\ \text{not having} \\ \text{information} \end{bmatrix}.$$

Thus

$$f'(t) = k \cdot [100 - f(t)] \qquad (1)$$

for some constant k. Equivalently,

$$y' = k(100 - y). \qquad (2)$$

Before we can sketch a solution, we must determine whether k is positive or negative. Clearly, f(t) is increasing, so f'(t) is positive. Also, f(t) is not more than 100 percent, so 100 - f(t) is also positive. From this it follows that k is positive.

From (2) we must sketch

$$z = k(100 - y) = 100k - ky,$$

where k is positive. The graph of this equation is a straight line with negative slope. Also, when y = 100 we have z = 0. See Fig. 1(a) below. We are interested in the solution where y(0) = 1. Using the method of Section 10.4, we obtain the curve in Fig. 1(b).

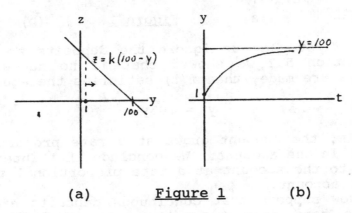

(a)   Figure 1   (b)

7. Let y = f(t) be the amount of substance A at time t. Then f'(t) is the rate at which substance A is converted into substance B. We are told that this rate is proportional to the square of the amount of A. Thus we have

$$f'(t) = k[f(t)]^2 \qquad (1)$$

where k is a constant of proportionality. Equivalently, we have

$$y' = ky^2. \tag{2}$$

Before we can sketch a solution, we must determine whether k is positive or negative. Clearly, f(t) is decreasing, so f'(t) is negative. Since $f^2(t)$ is never negative, k must be negative.

From (2) we must sketch,

$$z = ky^2,$$

where k is negative, so we have a parabola that opens down. See Fig. 1(a) below. We are interested in a solution where $|y(0)|$ is some positive number (representing the initial amount of substance A). Using the method of Section 10.4, we obtain the curve in Fig. 1(b).

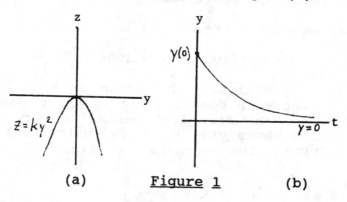

(a)   **Figure 1**   (b)

13. **a.** At first, let us ignore the deposits to the account. In Section 5.2 we showed that if no deposits or withdrawals are made, then f(t) satisfies the equation

$$y' = .05y.$$

That is, the account grows at a rate proportional to the amount in the account. We conclude that interest is being added to the account at a rate proportional to the amount in the account.

Now suppose that continuous deposits are being made to the same account at the rate of $10,000 per year. There are two influences on the way the amount of money in the account changes—the rate at which interest is added and the rate at which money is deposited. Let f(t) be the amount of money in the account at time t. Then the rate of change of f(t) is the net effect of these two influences. That is, f(t) satisfies the equation

$$f'(t) = .05f(t) + 10,000,$$

or equivalently,

$$y' = .05y + 10,000. \qquad\qquad (1)$$

At time $t = 0$, we assume there is no money in the account. Hence $y(0) = 0$.

b. We note that $y = \dfrac{-10,000}{.05} = -200,000$ is a constant solution, since it makes both sides of equation (1) equal zero. Assuming that $y \neq -200,000$, we proceed using separation of variables. We have

$$y' = .05(y + 200,000),$$

$$\int \frac{y'}{(y+200,000)} \frac{dy}{dt}\, dt = \int .05\, dt,$$

$$\int \frac{y'}{(y+200,000)}\, dy = \int .05\, dt,$$

$$\ln|y+200,000| = .05t + C, \quad C \text{ a constant},$$

$$|y+200,000| = e^{.05t+C} = e^{C} \cdot e^{.05t}.$$

If $y \neq -200,000$, then $y + 200,000$ is either positive or negative for all t. Thus

$$y = \pm e^{C} \cdot e^{.05t} - 200,000.$$

The general solution (including the constant solution) has the form

$$y = Ae^{.05t} - 200,000, \quad A \text{ a constant}.$$

Since we have $y(0) = 0$, we see that $A = 200,000$. Hence

$$y = 200,000e^{.05t} - 200,000$$
$$= 200,000(e^{.05t} - 1).$$

The amount in the account after 5 years is given by

$$y(t) = 200,000(e^{.25} - 1)$$
$$\approx 56,806 \text{ dollars}.$$

19. We must first sketch

$$z = -.0001q^{2}(1 - q).$$

Clearly $z = 0$ when $q = 0$ or $q = 1$. Writing $z = .0001q^{3} - .0001q^{2}$, we see that the graph is a cubic curve. To find relative extreme points we set $\frac{dz}{dq} = 0$.

$$.0003q^2 - .0002q = 0,$$

$$.0001q(3q - 2) = 0.$$

So $q = 0$, or $3q - 2 = 0$ and $q = 2/3$. Since $\frac{d^2z}{dq^2} = .0006q$ $- .0002 = .0002(3q-1)$, we see that $\frac{d^2z}{dq^2}$ is negative at $q = 0$ and positive at $q = 2/3$. Here is a rough sketch of the graph.

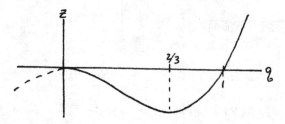

We are asked to find a solution when $q(0)$ is close to but slightly less than 1, so we assume $q(0) > 2/3$. Using the method of Section 10.4, we obtain

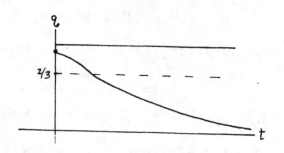

Review of Chapter 10

Except for one or two differential equations in Section 10.1, all of the equations have been *first-order* equations that involve only y′, y, and perhaps some functions of t. We have three methods for analyzing these equations: (1) separation of variables to find explicit solutions; (2) Euler's method to find an approximate solution; and (3) qualitative analysis of the equation. The chart below lists the methods that may be applied to first-order differential equations.

Form of the Equation	Methods of Solution
$y' = g(t,y)$	Euler's method
$y' = p(t)q(y)$	Euler's method, Separation of Variables
$y' = g(y)$	Euler's method, Separation of Variables, Qualitative Analysis

On an exam, your instructor might ask you to use more than one method on the same differential equation. This chart will help you anticipate what kinds of questions might be asked.

Chapter 10: Supplementary Exercises

1. We use separation of variables to obtain

$$\int y^2 \frac{dy}{dt}\, dt = \int (4t^3 - 3t^2 + 2)\, dt,$$

$$\int y^2 \, dy = \int (4t^3 - 3t^2 + 2)\, dt,$$

$$\tfrac{1}{3} y^3 = t^4 - t^3 + 2t + c_1,$$

$$y^3 = 3(t^4 - t^3 + 2t + c_1),$$

$$y^3 = 3(t^4 - t^3 + 2t) + C, \quad C \text{ a constant.}$$

Hence

$$y = \sqrt[3]{3t^4 - 3t^3 + 6t + C}.$$

7. We use separation of variables to obtain

$$yy' = 6t^2 - t,$$

$$\int y\frac{dy}{dt}\, dt = \int (6t^2 - t)\, dt,$$

$$\int y\, dy = \int (6t^2 - t)\, dt,$$

$$\frac{1}{2}y^2 = 2t^3 - \frac{1}{2}t^2 + C_1,$$

$$y^2 = 2(2t^3 - \frac{1}{2}t^2) + C, \ C \text{ a constant.}$$

Hence

$$y = \pm\sqrt{4t^3 - t^2 + C}.$$

Since $y(0) = 7$, we need the positive square root. (If $y(0)$ were negative here, we would use the negative square root.) Since $y(0) = \sqrt{C} = 7$, we have $C = 49$. Therefore,

$$y = \sqrt{4t^3 - t^2 + 49}.$$

13. Here $g(t,y) = .1y(20-y)$, $a = 0$, $b = 3$, $y_0 = 2$, and $h = \frac{3-0}{6}$ $= .5$. We have:

$$t_0 = 0, \quad y_0 = 2, \ g(0,2) = .2(20-2) = 3.6,$$

$$t_1 = .5, \quad y_1 = 2 + (3.6)(.5) = 3.8,$$
$$g(\tfrac{1}{2},3.8) = .38(20-3.8) = 6.16,$$

$$t_2 = 1, \quad y_2 = 3.8 + (6.16)(.5) = 6.9,$$
$$g(1,6.9) = .69(20-6.9) = 9.04,$$

$$t_3 = 1.5, \quad y_3 = 6.9 + (9.04)(.5) = 11.4,$$
$$g(3/2,11.4) = 1.14(20-11.4) = 9.8,$$

$$t_4 = 2, \quad y_4 = 11.4 + (9.8)(.5) = 16.3,$$
$$g(2,16.3) = 1.63(20-16.3) = 6.03,$$

$$t_5 = 2.5, \quad y_5 = 16.3 + (6.03)(.5) = 19.3,$$
$$g(5/2,19.3) = 1.93(20-19.3) = 1.35,$$

$$t_6 = 3, \quad y_6 = 19.3 + (1.35)(.5) = 19.98.$$

The polygonal path connecting the points $(t_0,y_0),\ldots,$ (t_6,y_6) is shown in the answer section of the text.

19. From $y' = \ln y$ we obtain the function $z = \ln y$, whose graph is shown in Fig. 1a. Thus $g(y)$ has a zero at $y = 1$. Therefore, the constant function $y = 1$ is a solution of the differential equation.

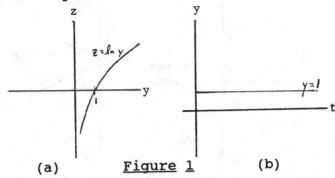

(a)   Figure 1   (b)

To sketch the solution corresponding to $y(0) = 2$, we observe that when $y = 2$, $g(y)$ is positive. Hence the solution is increasing as it leaves the initial point. As y moves to the right, the z-coordinate gets more positive, and hence the solution is concave up, and will continue to increase without bound. See Figure 2.

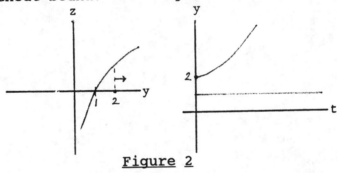

Figure 2

25. a. Let $N = f(t)$ be the city's population at time t. Then $f'(t)$ is the rate of change of the population of the city at time t. Since the birth rate is 3.5%, and the death rate is 2%, and 3000 people leave the city each year we have

$$f'(t) = .035f(t) - .02f(t) - 3000$$
$$= .015f(t) - 3000,$$

or equivalently,

$$N' = .015N - 3000.$$

b. We seek a constant function N which satisfies the differential equation $N' = .015N - 3000 = .015(N - 200,000)$. Clearly the constant function $N = 200,000$ satisfies the equation since it makes both sides equal to zero. However, it is unlikely that a city could maintain this constant population in practice, for the graph indicates that the constant solution is unstable.

We note that $z = .015N - 3000$ is the graph of a line with positive slope, and with a zero at $N = 200,000$. Using the method of Section 10.4, we obtain

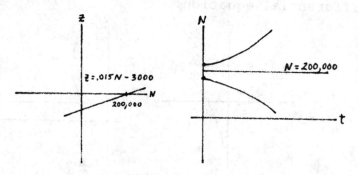

CHAPTER 11

PROBABILITY
AND CALCULUS

11.1 Discrete Random Variables

Although this section is provided mainly as a background for the sections that follow, there are interesting and important problems here. The main concepts are: relative frequency table, probability, expected value, variance, and (discrete) random variable. The probability density histogram on page 659 will be used in Section 11.2 to motivate the definition of a probability density function.

1. There are two possible outcomes, 0 and 1, with probabilities $\frac{1}{5}$ and $\frac{4}{5}$, respectively. Thus we have

$$E(X) = 0 \cdot \frac{1}{5} + 1 \cdot \frac{4}{5} = \frac{4}{5}.$$

$$Var(X) = \left(0 - \frac{4}{5}\right)^2 \cdot \frac{1}{5} + \left(1 - \frac{4}{5}\right)^2 \cdot \frac{4}{5} = .128 + .032 = .16.$$

Standard deviation of $X = \sqrt{.16} = .4$.

7. a. Recall that the area of a circle of radius r is given by [Area] $= \pi r^2$. Thus the area of a circle of radius 1 is π, and the area of a circle of radius $\frac{1}{2}$ is $\pi\left(\frac{1}{2}\right)^2 = \frac{1}{4}\pi$. To find the percentage of points lying in the circle of radius $\frac{1}{2}$, we set up the ratio

$$\frac{\left[\text{Area of circle of radius } \frac{1}{2}\right]}{\left[\text{Area of circle of radius 1}\right]} = \frac{\frac{1}{4}\pi}{\pi} = \frac{1}{4} = .25 = 25\%.$$

Thus 25% of the points lie within $\frac{1}{2}$ unit of the center.

b. The area of a circle of radius c is: [Area] = πc^2. To find the percentage of points lying within this circle we set up the ratio

$$\frac{[\text{Area of circle of radius c}]}{[\text{Area of circle of radius 1}]} = \frac{\pi c^2}{\pi} = c^2 = 100c^2\%.$$

11.2 Continuous Random Variables

We recommend that you study the theoretical parts of Sections 6.1 and 6.3 (particularly the fundamental theorem of calculus) as background for the discussion here of the cumulative distribution function F(x) and the probability density function f(x). Given either function, you should be able to find the other with no difficulty. Observe that F'(x) = f(x) and that F(x) is the unique antiderivative of f(x) for which F(A) = 0, when X is a random variable on A ≤ x ≤ B. You should also be able to use either F(x) or f(x) to compute various probabilities of the form Pr(a ≤ X ≤ b).

Pay attention to Example 6 because it provides a nice connection with earlier material on improper integrals and it introduces ideas that are needed in Section 11.4. For these reasons, questions similar to Example 6 often appear on exams.

1. Clearly f(x) ≥ 0, since $\frac{1}{18}x$ ≥ 0 when 0 ≤ x ≤ 6. Thus Property I is satisfied. For Property II we check that

$$\int_0^6 f(x)\,dx = \int_0^6 \frac{1}{18}x\,dx = \frac{1}{36}x^2\Big|_0^6 = \frac{1}{36}(6)^2 - 0 = 1.$$

Thus Property II is also satisfied, and f(x) is indeed a probability density function.

7. In order to satisfy Property I, we require that k ≥ 0, for if k < 0 and 1 ≤ x ≤ 3, then f(x) = kx < 0. For Property II we set $\int_1^3 kx\,dx = 1$, and solve for k.

$$\int_1^3 kx\,dx = \frac{k}{2}x^2\Big|_1^3 = \frac{k}{2}(9) - \frac{k}{2} = 4k = 1.$$

Therefore the value k = $\frac{1}{4}$ satisfies both Properties I and II, and f(x) = $\frac{1}{4}x$, 1 ≤ x ≤ 3, is a probability density function.

13. The set of possible values of X is determined by the domain of the density function f(x). In this problem the domain is 0 ≤ x ≤ 4, so the values of X range between 0 and 4. Since X can't be less than 0, the probability that X is less than or equal to 1 is the same as the probability that X is between 0 and 1. Thus Pr(X ≤ 1) is an abbreviation for Pr(0 ≤ X ≤ 1). Similarly, since X cannot be larger than 4, Pr(3.5 ≤ X) is an abbreviation for Pr(3.5 ≤ X ≤ 4).

a. Pr(X ≤ 1) is represented by the area under the graph of f(x) = $\frac{1}{8}$x where 0 ≤ x ≤ 1.

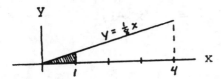

b. Pr(2 ≤ X ≤ 2.5) is represented by the area under the graph of f(x) = $\frac{1}{8}$x where 2 ≤ x ≤ 2.5.

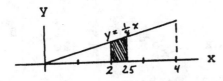

c. Pr(3.5 ≤ X) is represented by the area under the graph of f(x) = $\frac{1}{8}$x where 3.5 ≤ x ≤ 4.

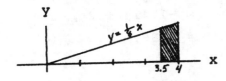

19. The probability that the lifetime X of a battery is *at least* 35 hours is given by Pr(35 ≤ X). However, we are told in this problem that the values of X range between 30 and 50. So Pr(35 ≤ X) = Pr(35 ≤ X ≤ 50), and

$$\text{Pr}(35 \leq X) = \int_{35}^{50} \frac{1}{20}\, dx = \frac{1}{20}x \Big|_{35}^{50} = \frac{50}{20} - \frac{35}{20} = \frac{15}{20} = \frac{3}{4}.$$

Thus the probability is $\frac{3}{4}$ that the battery will last longer than 35 hours.

25. a. $Pr(2 \leq X \leq 3) = \int_2^3 \frac{1}{21}x^2 \, dx = \frac{1}{63}x^3 \Big|_2^3 = \frac{27}{63} - \frac{8}{63} = \frac{19}{63}$.

b. Since $F(x)$ is an antiderivative of $f(x) = \frac{1}{21}x^2$, we have

$$F(x) = \int \frac{1}{21}x^2 \, dx = \frac{1}{63}x^3 + C.$$

Since X has values greater than or equal to 1, we have $F(1) = 0$. Setting $F(1) = \frac{1}{63} + C = 0$, we find that $C = -\frac{1}{63}$, so

$$F(x) = \frac{1}{63}x^3 - \frac{1}{63} = \frac{1}{63}(x^3-1).$$

c. $Pr(2 \leq X \leq 3) = F(3) - F(2)$

$$= \frac{1}{63}(3^3-1) - \frac{1}{63}(2^3-1) = \frac{26}{63} - \frac{7}{63}$$

$$= \frac{19}{63}.$$

31. $Pr(0 \leq X \leq 5) = \int_0^5 2ke^{-kx} \, dx = \frac{2k}{-k}e^{-kx} \Big|_0^5 = -2e^{-kx} \Big|_0^5.$

Since $k = (\ln 2)/10 \approx .0693$, we have

$$Pr(0 \leq X \leq 5) = -2e^{-.0693x} \Big|_0^5 = -2e^{(-.0693)(5)} - [-2e^0]$$

$$\approx -1.414 + 2$$

$$= .586.$$

Alternate calculation:

$$-2e^{-kx} \Big|_0^5 = -2e^{-k(5)} - (-2e^0) = -2e^{-5(\ln 2)/10} + 2.$$

Observe that

$$-2e^{-5(\ln 2)/10} = -2e^{-(1/2)\ln 2} = -2e^{\ln 2^{-1/2}}$$

$$= -2 \cdot 2^{-1/2} = -2^{1/2} = -\sqrt{2}.$$

Thus

$$Pr(0 \leq X \leq 5) = -\sqrt{2} + 2 \approx -1.414 + 2 = .586.$$

37. a. Clearly $f(x) = 4x^{-5} \geq 0$ for all values of x greater than or equal to 1, so Property I is satisfied. For Property II we check that $\int_1^\infty 4x^{-5} \, dx = 1$. If $b \geq 1$, then

$$\int_1^b 4x^{-5} \, dx = -x^{-4}\Big|_1^b = -(b)^{-4} - [-(1)^{-4}] = 1 - \frac{1}{b^4}.$$

Since as $b \to \infty$, the number b^4 gets arbitrarily large, $\frac{1}{b^4}$ approaches zero. Thus

$$\int_1^\infty 4x^{-5} \, dx = \lim_{b \to \infty} \int_1^b 4x^{-5} \, dx = \lim_{b \to \infty} \left(1 - \frac{1}{b^4}\right) = 1.$$

b. Since $F(x)$ is an antiderivative of $f(x) = 4x^{-5}$, we have

$$F(x) = \int 4x^{-5} \, dx = -x^{-4} + C.$$

Since X has values greater than or equal to 1, we have $F(1) = 0$. Setting $F(1) = -1 + C = 0$, we find that $C = 1$, so

$$F(x) = -x^{-4} + 1 = 1 - x^{-4}.$$

c. $Pr(1 \leq X \leq 2) = F(2) - F(1) = \frac{15}{16} - 0 = \frac{15}{16}.$ Observe that $Pr(1 \leq X \leq 2) + Pr(2 \leq X) = 1$, since <u>all</u> values of X are greater than or equal to 1. Hence

$$Pr(2 \leq X) = 1 - Pr(1 \leq X \leq 2) = 1 - \frac{15}{16} = \frac{1}{16}.$$

Helpful Hint: Decide now what you would do on an exam if you were given Exercise 37 with no mention of parts (a), (b), and $Pr(1 \leq x \leq 2)$. Certainly there would be no need to find the cumulative distribution function. Would you want to compute $Pr(1 \leq X \leq 2)$? Compute (a) and (b) below, and decide which approach is easier for you.

(a) $Pr(2 \leq X) = 1 - Pr(1 \leq X \leq 2) = 1 - \int_1^2 4x^{-5} \, dx.$

(b) $Pr(2 \leq X) = \int_2^\infty 4x^{-5} \, dx.$

11.3 Expected Value and Variance

The expected value of a random variable X is easier to understand and compute than the variance. However, both concepts are important in statistics. An explanation of how E(X) and Var(X) are related to the graph of the probability density function for X will be given in Section 11.4, in the discussion of normal random variables. Check with your instructor to see if you need to know the definition of Var(X), or if the alternate formula for Var(X) will suffice. The alternate formula is easier to compute.

Calculations of E(X) and Var(X) sometimes seem to require integration by parts, but this can be avoided when the probability density function is a polynomial. Observe how the integrals $\int_0^1 xf(x)\,dx$ and $\int_0^1 x^2 f(x)\,dx$ in Practice Problem 2 are simplified by expanding $f(x)$ so that $xf(x)$ and $x^2 f(x)$ are simple polynomials that are easily integrated.

1. We have

$$E(X) = \int_0^6 x \cdot \frac{1}{18}x\,dx = \int_0^6 \frac{1}{18}x^2\,dx = \frac{1}{54}x^3 \Big|_0^6 = 4.$$

To compute the alternate formula for Var(X), we compute

$$\int_0^6 x^2 \cdot \frac{1}{18}x\,dx = \int_0^6 \frac{1}{18}x^3\,dx = \frac{1}{72}x^4 \Big|_0^6 = \frac{1296}{72} = 18.$$

Therefore

$$Var(X) = \int_0^6 x^2 \cdot \frac{1}{18}x\,dx - E(X)^2 = 18 - (4)^2 = 2.$$

7. We have

$$E(X) = \int_0^1 x \cdot 12x(1-x)^2\,dx = \int_0^1 12x^2(1-2x+x^2)\,dx$$

$$= \int_0^1 (12x^2 - 24x^3 + 12x^4)\,dx$$

$$= (4x^3 - 6x^4 + \frac{12}{5}x^5) \Big|_0^1 = \frac{2}{5}.$$

11-6

$$Var(X) = \int_0^1 x^2 \cdot 12x(1-x)^2 \, dx - E(X)^2$$

$$= \int_0^1 (12x^3 - 24x^4 + 12x^5) \, dx - \left(\frac{2}{5}\right)^2$$

$$= (3x^4 - \frac{24}{5}x^5 + 2x^6)\Big|_0^1 - \frac{4}{25}$$

$$= \frac{1}{5} - \frac{4}{25} = \frac{1}{25}.$$

13. The average time is the expected value of X.

$$E(X) = \int_0^{12} \frac{1}{72}x^2 \, dx = \frac{1}{216}x^3\Big|_0^{12} = \frac{1728}{216} = 8.$$

Therefore, the average time spent reading the editorial page is 8 minutes.

19. We set $\int_0^M \frac{1}{18}x \, dx = \frac{1}{2}$ and solve for M.

$$\int_0^M \frac{1}{18}x \, dx = \frac{1}{36}x^2\Big|_0^M = \frac{1}{36}M^2 = \frac{1}{2}.$$

So $M^2 = 18$, and $M = \sqrt{18} = 3\sqrt{2}$. We note that $-3\sqrt{2}$ is not a solution since $0 \le M \le 6$.

25. We have $E(X) = \int_A^B xf(x) \, dx$, so we integrate by parts, setting

$$h(x) = x, \quad f(x) \quad [= g(x) \text{ in integ. by parts formula}]$$

$$h'(x) = 1, \quad F(x) \quad [= \text{antiderivative of } f(x)]$$

where $f(x)$ is any probability density function on $A \le x \le B$ and $F(x)$ is the cumulative distribution function corresponding to $f(x)$. So we have

$$E(X) = \int_A^B xf(x) \, dx = xF(x)\Big|_A^B - \int_A^B F(x) \, dx$$

11-7

$$= BF(B) - AF(A) - \int_A^B F(x) \, dx$$

$$= B - \int_A^B F(x) \, dx,$$

since $F(B) = 1$ and $F(A) = 0$.

11.4 Exponential and Normal Random Variables

This section deserves your attention because exponential and normal random variables arise so often in applications. In a situation when an exponential density function ke^{-kx} is appropriate for a random variable X, the value of $E(X)$ is often estimated from experimental data, and then the constant k is obtained from the relation

$$E(X) = \frac{1}{k}, \quad \text{or equivalently,} \quad k = \frac{1}{E(X)}.$$

You need to know this relation in order work Exercises 5—14.

The text's discussion of arbitrary normal random variables provides an opportunity to review changes of variable in a definite integral. Check with your instructor about how much detail you should show when you make a substitution $z = (x - \mu)/\sigma$.

1. Here $k = 3$, so we have

$$E(X) = \frac{1}{3}, \quad \text{and Var}(X) = \frac{1}{9}.$$

7. The mean is given by $E(X) = \frac{1}{k} = 3$. Hence $k = \frac{1}{3}$, and the probability density function is $f(x) = \frac{1}{3}e^{-(1/3)x}$. The probability that a customer is served in less than 2 minutes is given by

$$Pr(0 \leq X \leq 2) = \int_0^2 \frac{1}{3}e^{-(1/3)x} \, dx = -e^{-(1/3)x}\Big|_0^2$$

$$= -e^{-2/3} - (-e^0)$$

$$= 1 - e^{-2/3}.$$

13. The average life span (or mean) is given by $E(X) = \frac{1}{k} = 72$. Hence $k = \frac{1}{72}$, and the probability density function is $f(x) = \frac{1}{72}e^{-(1/72)x}$. The probability that a component lasts for more than 24 months is given by

$$Pr(24 \leq X) = \int_{24}^{\infty} \frac{1}{72}e^{-(1/72)x}\, dx$$

$$= \lim_{b \to \infty} \int_{24}^{b} \frac{1}{72}e^{-(1/72)x}\, dx$$

$$= \lim_{b \to \infty} \left[-e^{-(1/72)x} \Big|_{24}^{b} \right]$$

$$= \lim_{b \to \infty} \left[e^{-1/3} - e^{-b/72} \right].$$

We note that as $b \to \infty$, the number $-\frac{b}{72}$ approaches $-\infty$, so $e^{-b/72}$ approaches zero. Therefore

$$Pr(24 \leq X) = \lim_{b \to \infty} \left[e^{-1/3} - e^{-b/72} \right] = e^{-1/3}.$$

19. To find a relative maximum, we first set $f'(x) = 0$ and solve for x.

$$f'(x) = -xe^{-x^2/2} = 0.$$

Since $e^{-x^2/2}$ is strictly positive for all values of x, we see that $f'(x) = 0$ if and only if $x = 0$. To check that $f(x)$ has a maximum at $x = 0$, we check $f''(0)$.

$$f''(x) = x^2 e^{-x^2/2} - e^{-x^2/2}$$

$$f''(0) = 0 - 1 = -1 < 0.$$

Thus $f(x)$ is indeed concave down at $x = 0$. Therefore $f(x)$ has a relative maximum at $x = 0$.

25. a. We have $\mu = 6$ and $\sigma = \frac{1}{2}$. Note that $\frac{x-\mu}{\sigma} = \frac{x-6}{1/2} = 2x - 12$ and $\frac{1}{\sigma\sqrt{2\pi}} = \frac{1}{(1/2)\sqrt{2\pi}} = \frac{2}{\sqrt{2\pi}}$. Thus the normal density function for X, the gestation period, is $f(x) = \frac{2}{\sqrt{2\pi}}e^{-(1/2)(2x-12)^2}$. So we have

$$\Pr(6 \le X \le 7) = \int_6^7 \frac{2}{\sqrt{2\pi}}e^{-(1/2)(2x-12)^2}\,dx.$$

Using the substitution $z = 2x - 12$, $dz = 2\,dx$, we see that if $x = 6$, then $z = 0$, and if $x = 7$, then $z = 2$. So

$$\int_6^7 \frac{2}{\sqrt{2\pi}}e^{-(1/2)(2x-12)^2}\,dx = \int_0^2 \frac{1}{\sqrt{2\pi}}e^{-(1/2)z^2}\,dz.$$

The value of this integral is the area under the standard normal curve from 0 to 2. Thus

$$\Pr(6 \le X \le 7) = A(2) = .4772.$$

Therefore, about 47.72% of births occur after a gestation period of between 6 and 7 months.

b. Proceeding as in part (a), we use the same substitution to evaluate $\int_5^6 \frac{2}{\sqrt{2\pi}}e^{-(1/2)(2x-12)^2}\,dx$, noting that if $x = 5$, then $z = 2x - 12 = -2$, and if $x = 6$, then $z = 0$. Therefore,

$$\Pr(5 \le X \le 6) = \int_{-2}^0 \frac{2}{\sqrt{2\pi}}e^{-(1/2)z^2}\,dz$$

$$= A(-2) = A(2) = .4772.$$

Therefore, about 47.72% of births occur after a gestation period of between 5 and 6 months.

31. We have the normal density function

$$f(x) = \frac{1}{(.8)\sqrt{2\pi}}e^{-(1/2)[(x-18.2)/.8]^2}.$$

Let X be the diameter of a bolt selected at random from the supply of bolts. We have

$$Pr(20 \leq X) = \int_{20}^{\infty} \frac{1}{(.8)\sqrt{2\pi}} e^{-(1/2)[(x-18.2)/.8]^2} dx.$$

Using the substitution $z = (x - 18.2)/.8$, $dz = (1/.8)\,dx$, we note that if $x \to \infty$, then $z \to \infty$ too, and if $x = 20$, then $z = 1.8/.8 = 2.25$. Hence

$$Pr(20 \leq X) = \int_{2.25}^{\infty} \frac{1}{\sqrt{2\pi}} e^{-(1/2)z^2} dz.$$

The value of this integral is the area under the standard normal curve from 2.25 to ∞. If we consider areas under the standard normal curve, we see that

$$\begin{bmatrix} \text{area to} \\ \text{right of } 2.25 \end{bmatrix} = \begin{bmatrix} \text{area to} \\ \text{right of } 0 \end{bmatrix} - \begin{bmatrix} \text{area between} \\ 0 \text{ and } 2.25 \end{bmatrix}.$$

That is,

$$Pr(20 \leq X) = .5 \qquad - \qquad A(2.25)$$
$$= .5 - .4878$$
$$= .0122.$$

Therefore, about 1.22% of the bolts will be discarded.

Review of Chapter 11

Study the Chapter Checklist and determine which terms and formulas you must memorize. Summarized below are the main skills you should have, listed by the various types of random variables.

   a. *Discrete random variable.* Compute $E(X)$, $Var(X)$, and the standard deviation of X.

   b. *Continuous random variable on a finite interval.* Compute $E(X)$, $Var(X)$; test if $f(x)$ is a probability density function; given one of $f(x)$ and $F(x)$, find the other; use $f(x)$ or $F(x)$ to compute probabilities. Don't forget that $Pr(X \leq b)$ stands for $Pr(A \leq X \leq b)$, etc. Harder questions are found in Exercises 32–36.

   c. *Continuous random variable on an infinite interval.* A typical probability density function is $f(x) = kx^{-(k+1)}$ for $x \geq 1$. Same skills as for (**b**), plus use the formula $Pr(X \geq b) = 1 - Pr(X \leq b)$ to simplify calculations.

    d. *Exponential random variable.* The probability density function is $f(x) = ke^{-kx}$ for $x \geq 0$. Show that $f(x)$ *is* a probability density function. Compute various probabilities and the cumulative distribution function. You probably will not have to compute $E(X)$ and $Var(X)$, unless your instructor gives you the following limits to learn:

$$\lim_{b \to \infty} be^{-kb} = 0 \quad \text{and} \quad \lim_{b \to \infty} b^2 e^{-kb} = 0 \quad \text{for } k > 0.$$

However, you should know that $E(X) = 1/k$ and $Var(X) = 1/k^2$.

    e. *Normal random variables.* Use a table to compute probabilities for a standard normal random variable, and make a change of variable to compute probabilties for an arbitrary normal random variable.

Chapter 11: Supplementary Exercises

1.   **a.** $Pr(X \leq 1) = \displaystyle\int_0^1 \frac{3}{8}x^2 \, dx = \frac{1}{8}x^3 \Big|_0^1 = \frac{1}{8} = .125.$

    b. $Pr(1 \leq X \leq 1.5) = \displaystyle\int_1^{1.5} \frac{3}{8}x^2 \, dx = \frac{1}{8}x^3 \Big|_1^{1.5}$

$$= .4219 - .125 = .2969.$$

7.   **a.** From the definition of expected value, we have

$$E(X) = 1(.599) + 11(.401) = 5.01.$$

This can be interpreted as saying that if a large number of samples are tested in batches of ten samples per batch, then the average number of tests per batch will be close to 5.01.

b. The 200 samples will be divided into 20 batches of ten samples. In part (a) we found that, on the average, 5.01 tests must be run on each batch. So the laboratory should expect to run about $(20)(5.01) \approx 100$ tests altogether.

13. **a.** We must find the value of k that satisfies the equation

$$\int_5^{25} kx\, dx = 1.$$

We compute

$$\int_5^{25} kx\, dx = \frac{k}{2}x^2 \Big|_5^{25} = \frac{k}{2}625 - \frac{k}{2}25 = 300k = 1.$$

Therefore $k = \frac{1}{300}$.

b. $Pr(20 \le X \le 25) = \int_{20}^{25} \frac{1}{300}x\, dx = \frac{1}{600}x^2 \Big|_{20}^{25}$

$$= \frac{1}{600}625 - \frac{1}{600}400 = \frac{1}{600}(625 - 400) = .375.$$

c. $E(X) = \int_5^{25} \frac{1}{300}x^2\, dx = \frac{1}{900}x^3 \Big|_5^{25} = \frac{1}{900}(15,625 - 125)$

$$= 17.222.$$

Therefore, the mean annual income is \$17,222.

19. We first show that $Pr(X \le 4) = 1 - e^{-4k}$. We have

$$Pr(X \le 4) = \int_0^4 ke^{-kx}\, dx = -e^{-kx} \Big|_0^4 = 1 - e^{-4k}.$$

Since $Pr(X \le 4) = .75$, we have

$$1 - e^{-4k} = .75$$

$$e^{-4k} = .25$$

$$-4k = \ln .25 \approx -1.386$$

$$k = \frac{-1.386}{-4} \approx .35.$$

25. We know that $Pr(a \le Z)$ is the area under the standard normal curve to the right of a. Since $Pr(a \le Z) = .4$, and $Pr(0 \le Z) = .5$, we conclude that a lies to the right of zero. That is, a is positive. Thus

$$\begin{bmatrix} \text{area to} \\ \text{right of a} \end{bmatrix} = \begin{bmatrix} \text{area to} \\ \text{right of 0} \end{bmatrix} - \begin{bmatrix} \text{area between} \\ 0 \text{ and a} \end{bmatrix}$$

$$\Pr(a \leq Z) = \Pr(0 \leq Z) - \Pr(0 \leq Z \leq a)$$

$$.4 = .5 - \Pr(0 \leq Z \leq a).$$

So we have $\Pr(0 \leq Z \leq a) = A(a) = .1$. From the $A(z)$ table, we look for the z that makes $A(z)$ as close to .1 as possible. Since $A(.25) = .0987 \approx .1000$, we conclude that $a \approx .25$.

CHAPTER 12

TAYLOR POLYNOMIALS AND INFINITE SERIES

12.1 Taylor Polynomials

Taylor polynomials are often used in applications to approximate more complex functions, and they have an important connection with the Taylor series, to be discussed in Section 12.6.

The phrase "at x = 0" in "Taylor polynomial of f(x) at x = 0" indicates that the coefficients of the polynomial are computed by evaluating f(x) and its derivatives at x = 0. The phrase "at x = 0" does <u>not</u> mean that the x in the polynomial must be zero. However, in an application, the values of x are usually taken to be close to zero. In such cases, the remainder theorem is useful for estimating how close the values of a Taylor polynomial are to the values of the function f(x).

1. $f(x) = \sin x,$ $f(0) = 0,$

         $f'(x) = \cos x,$ $f'(0) = 1,$

         $f''(x) = -\sin x,$ $f''(0) = 0,$

         $f^{(3)}(x) = -\cos x,$ $f^{(3)}(0) = -1.$

Therefore, the third Taylor polynomial at x = 0 is

$$p_3(x) = 0 + \frac{1}{1!}x + \frac{0}{2!}x^2 + \frac{-1}{3!}x^3$$

$$= x - \frac{1}{6}x^3.$$

7. $f(x) = xe^{3x},$ $f(0) = 0,$

         $f'(x) = 3xe^{3x} + e^{3x},$ $f'(0) = 1,$

         $f''(x) = 3e^{3x} + 3e^{3x} + 9xe^{3x},$ $f''(0) = 6,$

$$f^{(3)}(x) = 9e^{3x} + \underbrace{9e^{3x} + 27xe^{3x}} + 9e^{3x}, \quad f^{(3)}(0) = 27.$$

from product
rule

Therefore,

$$p_3(x) = 0 + \frac{1}{1!}x + \frac{6}{2!}x^2 + \frac{27}{3!}x^3$$

$$= x + 3x^2 + \frac{9}{2}x^3.$$

13. Since $f(x) = f'(x) = f''(x) = \ldots = f^{(n)}(x) = e^x$, we have

$$f(0) = f'(0) = f''(0) = \ldots = f^{(n)}(0) = e^0 = 1.$$

Therefore, the nth Taylor polynomial for $f(x) = e^x$ at $x = 0$ is:

$$p_n(x) = 1 + \frac{1}{1!}x + \frac{1}{2!}x^2 + \frac{1}{3!}x^3 + \ldots + \frac{1}{n!}x^n$$

$$= 1 + x + \frac{1}{2}x^2 + \frac{1}{3!}x^3 + \ldots + \frac{1}{n!}x^n.$$

19.

$$\begin{array}{ll} f(x) = \cos x, & f(\pi) = -1, \\ f'(x) = -\sin x, & f'(\pi) = 0, \\ f''(x) = -\cos x, & f''(\pi) = 1, \\ f^{(3)}(x) = \sin x, & f^{(3)}(\pi) = 0, \\ f^{(4)}(x) = \cos x, & f^{(4)}(\pi) = -1. \end{array}$$

Thus the third Taylor polynomial at $x = \pi$ is

$$p_3(x) = -1 + \frac{0}{1!}(x-\pi) + \frac{1}{2!}(x-\pi)^2 + \frac{0}{3!}(x-\pi)^3$$

$$= -1 + \frac{1}{2}(x-\pi)^2,$$

and the fourth Taylor polynomial is

$$p_4(x) = -1 + \frac{0}{1!}(x-\pi) + \frac{1}{2!}(x-\pi)^2 + \frac{0}{3!}(x-\pi)^3 + \frac{-1}{4!}(x-\pi)^4$$

$$= -1 + \frac{1}{2}(x-\pi)^2 - \frac{1}{24}(x-\pi)^4.$$

Notice that $p_3(x)$ is a polynomial of degree 2, not degree 3, because $f^{(3)}(\pi) = 0$. See Practice Problem 1.

25.

$$f(z) = \frac{1}{\sqrt{2\pi}}e^{-z^2/2}, \qquad\qquad f(0) = \frac{1}{\sqrt{2\pi}},$$

$$f'(z) = \frac{-z}{\sqrt{2\pi}}e^{-z^2/2}, \qquad\qquad f'(0) = 0,$$

$$f''(z) = \frac{z^2}{\sqrt{2\pi}}e^{-z^2/2} - \frac{1}{\sqrt{2\pi}}e^{-z^2/2}, \quad f''(0) = -\frac{1}{\sqrt{2\pi}}.$$

Therefore, the second Taylor polynomial is

$$p_2(z) = \frac{1}{\sqrt{2\pi}} + \frac{0}{1!}z + \frac{-1/\sqrt{2\pi}}{2!}z^2$$

$$= \frac{1}{\sqrt{2\pi}} - \frac{1}{2\sqrt{2\pi}}z^2.$$

31. We have $f(9) = 3$,

$$f'(x) = \frac{1}{2}x^{-1/2}, \qquad f'(9) = \frac{1}{2}\cdot\frac{1}{9^{1/2}} = \frac{1}{2}\cdot\frac{1}{3} = \frac{1}{6},$$

$$f''(x) = -\frac{1}{4}x^{-3/2}, \qquad f''(9) = -\frac{1}{4}\cdot\frac{1}{9^{3/2}} = -\frac{1}{4}\cdot\frac{1}{3^3} = -\frac{1}{108}.$$

The second Taylor polynomial at $x = 9$ is

$$p_2(x) = 3 + \frac{1}{6}(x-9) - \frac{1}{216}(x-9)^2.$$

a. Note that $f^{(3)}(x) = \frac{3}{8}x^{-5/2}$. Hence at $x = 9$,

$$R_2(x) = \frac{f^{(3)}(9)}{3!}(x-9)^3 = \frac{3}{8}c^{-5/2}\cdot\frac{1}{3!}(x-9)^3,$$

where c is between 9 and x.

b. If $c \geq 9$, then $c^{5/2} \geq 9^{5/2} = (9^{1/2})^5 = 3^5 = 243$, and $c^{-5/2} \leq 1/243$. Thus

$$|f^{(3)}(c)| = \frac{3}{8}c^{-5/2} \leq \frac{3}{8}\cdot\frac{1}{243} = \frac{1}{648}.$$

c. The error in using $p_2(9.3)$ as an approximation for $\sqrt{9.3}$ is not more than $|R_2(9.3)|$. From (a) and (b),

$$|R_2(9.3)| \leq \frac{1}{648}\cdot\frac{1}{3!}(9.3-9)^3 = \frac{1}{648}\cdot\frac{1}{6}\left(\frac{3}{10}\right)^3$$

$$= \frac{27}{648\cdot6} \times 10^{-3} = \frac{1}{144} \times 10^{-3}.$$

12.2 The Newton-Raphson Algorithm

A calculator is indispensable for this section. In order to memorize the Newton-Raphson algorithm, it is important to work several problems by hand using a calculator. If you have access to a microcomputer, you might enjoy comparing your work to the results obtained by the computer. Here is a BASIC program that works for Exercise 7.

```
10 LET X = 0
20 FOR J = 1 TO 20
30    LET X = X - (SIN(X) + X*X - 1)/(COS(X) + 2*X)
40 NEXT J
50 PRINT X
99 END
```

If you know BASIC, you'll probably think of several ways in which to improve this program.

1. $\sqrt{5}$ is a zero of the function $x^2 - 5$. Since $\sqrt{5}$ clearly lies between 2 and 3, let us take our initial approximation as $x_0 = 2$. Since $f'(x) = 2x$, we have:

$$x_1 = x_0 - \frac{x_0^2 - 5}{2x_0} = 2 - \frac{(2)^2 - 5}{2(2)}$$

$$= 2 - \frac{-1}{4} = \frac{9}{4} = 2.25,$$

$$x_2 = 2.25 - \frac{(2.25)^2 - 5}{2(2.25)}$$

$$= 2.25 - \frac{.0625}{4.5} = 2.2361,$$

$$x_3 = 2.2361 - \frac{(2.2361)^2 - 5}{2(2.2361)}$$

$$= 2.2361 - \frac{.00014}{4.4722} = 2.23607.$$

7. We have $f(x) = \sin x + x^2 - 1$, and $f'(x) = \cos x + 2x$. With $x_0 = 0$, we obtain:

$$x_1 = x_0 - \frac{\sin x_0 + x_0^2 - 1}{\cos x_0 + 2x_0} = 0 - \frac{\sin 0 + 0^2 - 1}{\cos 0 + 2(0)}$$

$$= 0 - \frac{-1}{1} = 1,$$

$$x_2 = 1 - \frac{\sin 1 + 1^2 - 1}{\cos(1) + 2(1)} = 1 - \frac{.84147}{2.54030} = .66875,$$

$$x_3 = .66875 - \frac{.62001 + .44723 - 1}{.78460 + 1.3375}$$

$$= .66875 - .03168 = .63707.$$

13. Let i be the monthly rate of interest. The present value of an amount A to be received in k months is $A(1+i)^{-k}$. Therefore, we must solve the following equation for i:

$$\begin{bmatrix} \text{amount of initial} \\ \text{investment} \end{bmatrix} = \begin{bmatrix} \text{sum of present} \\ \text{values of returns} \end{bmatrix}$$

$$500 = 100(1+i)^{-1} + 200(1+i)^{-2} + 300(1+i)^{-3}.$$

We multiply both sides by $(1+i)^3$, take all terms to the left, and obtain

$$500(1+i)^3 - 100(1+i)^2 - 200(1+i) - 300 = 0.$$

Let $x = 1+i$, and solve the resulting equation by the Newton-Raphson algorithm with $x_0 = 1.1$.

$$f(x) = 500x^3 - 100x^2 - 200x - 300 = 0,$$
$$f'(x) = 1500x^2 - 200x - 200.$$

$$x_1 = 1.1 - \frac{500(1.1)^3 - 100(1.1)^2 - 200(1.1) - 300}{1500(1.1)^2 - 200(1.1) - 200}$$

$$\approx 1.1 - .018$$

$$= 1.082,$$

$$x_2 = 1.082 - \frac{500(1.082)^3 - 100(1.082)^2 - 200(1.082) - 300}{1500(1.082)^2 - 200(1.082) - 200}$$

$$\approx 1.082 - (-.00008)$$

$$\approx 1.082.$$

Therefore, the solution is $x = 1.082$. Hence $i = .082$ and the investment had an internal rate of return of 8.2% per month.

19. Slope of the tangent line is 4 ($y = mx + b$), so $f'(3) = 4$, and $f(3) = 4(3) + 5 = 17$. Plugging these into the Newton-Raphson formula with $x_0 = 3$ as our initial guess:

$$x_1 = x_0 - \frac{f(x_0)}{f'(x_0)} = 3 - \frac{17}{4} = \frac{-5}{4}.$$

25. We have $f(x) = x^{1/3}$, and $f'(x) = \frac{1}{3}x^{-2/3} = \frac{1}{3x^{2/3}}$. With $x_0 = 1$ as our initial guess: $f(1) = 1$, $f'(1) = 1/3$. Thus by Newton's method,

$$x_1 = x_0 - \frac{f(x_0)}{f'(x_0)} = 1 - \frac{1}{1/3} = 1 - 3 = -2,$$

$$f(x_1) = f(-2) = -(2)^{1/3} = -1.260,$$

$$f'(x_1) = f'(-2) = -\frac{1}{3(2)^{2/3}} = -.210.$$

$$x_2 = -2 - \frac{1.26}{.21} = -2 - 6 = -8.$$

Thus, since from the diagram we can see that the root is x = 0, the iterates are diverging away from this value. Therefore Newton's method does not converge.

12.3 Infinite Series

Most students who take this calculus course will see an application of infinite series in some other course. More often than not, the application will involve a geometric series.

Make sure you completely master the material in this section. It will really help you to understand the concepts in the sections to follow.

1. $1 + \frac{1}{6} + \frac{1}{6^2} + \frac{1}{6^3} + \frac{1}{6^4} + \ldots = (\frac{1}{6})^0 + (\frac{1}{6})^1 + (\frac{1}{6})^2 + (\frac{1}{6})^3 + (\frac{1}{6})^4 + \ldots$

Thus $a = 1$, $r = \frac{1}{6}$, and the series converges to

$$\frac{a}{1-r} = \frac{1}{1-\frac{1}{6}} = \frac{1}{\frac{5}{6}} = \frac{6}{5}.$$

7. We find r by dividing any term by the preceding term. So

$$r = \frac{\frac{1}{5^4}}{\frac{1}{5}} = \frac{5}{5^4} = \frac{1}{5^3} = \frac{1}{125}.$$

The first term of the series is $a = \frac{1}{5}$, so the sum of the series is $\dfrac{\frac{1}{5}}{1 - \frac{1}{125}} = \frac{1}{5} \cdot \frac{125}{124} = \frac{25}{124}.$

13. We find r by dividing any term by the preceding term. So,

$$r = \frac{4}{5}.$$

The first term of the series is $a = 5$, so the sum of the series is $\dfrac{5}{1 - \frac{4}{5}} = 5 \cdot 5 = 25.$

19. The technique here is to find a rational number that represents .011<u>011</u>, and then to add 4 to this number. We have

$$.011\underline{011} = .011 + .000011 + .000000011 + \cdots$$

$$= \frac{11}{1000} + \frac{11}{1000^2} + \frac{11}{1000^3} + \cdots,$$

which is a geometric series with $a = \frac{11}{1000}$ and $r = \frac{1}{1000}$.

The sum of the series is: $\dfrac{\frac{11}{1000}}{1 - \frac{1}{1000}} = \frac{11}{1000} \cdot \frac{1000}{999} = \frac{11}{999}$. So

$$4.011\underline{011} = 4 + \frac{11}{999} = \frac{3996}{999} + \frac{11}{999} = \frac{4007}{999}.$$

25. a. We construct a sum of the present value of all future payments. That is, we express capital value as

$$100 + 100(1.01)^{-1} + 100(1.01)^{-2} + \cdots = \sum_{k=0}^{\infty} 100(1.01)^{-k}.$$

b. This is a geometric series. We have $a = 100$ and $r = (1.01)^{-1} = \frac{1}{1.01}$, (so $|r| < 1$). So the sum of the series is $\dfrac{100}{1 - \frac{1}{1.01}} = 100 \, \frac{1.01}{.01} = 10,100$. Thus the capital value of the perpetuity is $10,100.

31. At the end of day 1 the body has M mg present.

At end of day 2 the body has M mg obtained on day 2 plus $\frac{3}{4}$ of the previous day's dose, i.e., it has eliminated $\frac{1}{4}$ of the previous day's dose. Thus the total at the end of day 2 is $\left(M + \frac{3}{4}\right)$ mg still left in the body.

Similarly, the day 3 total is $M + \frac{3}{4}\left(M + \frac{3}{4}M\right)$ mg

$$= \left(M + \frac{3}{4}M + \left(\frac{3}{4}\right)^2 M\right) \text{ mg.}$$

Extrapolating, the total after many days is

$$M + \frac{3}{4}M + \left(\frac{3}{4}\right)^2 M + \left(\frac{3}{4}\right)^3 M + \cdots$$

$$= M\left[1 + \frac{3}{4} + \left(\frac{3}{4}\right)^2 + \left(\frac{3}{4}\right)^3 + \cdots\right]$$

$$= \sum_{k=0}^{N} M(3/4)^k, \quad N = \text{number of days}.$$

This is similar to a geometric series with $r = 3/4$ and $M = 1$. So since

$$\sum_{k=0}^{\infty} ar^k = \frac{a}{1-r}, \quad r < 1.$$

$$\sum_{k=0}^{N} M(3/4)^k \approx \frac{M}{1 - 3/4} = 4M.$$

Therefore, for 20 mg to be present immediately after a dose is given:

$$4M = 20, \quad \text{or} \quad M = 5 \text{ mg}.$$

37. $\displaystyle\sum_{j=1}^{\infty} 5^{-j} = \frac{1}{5} + \frac{1}{5^2} + \frac{1}{5^3} + \frac{1}{5^4} + \dots$

This is a geometric series with $a = \frac{1}{5}$ and $r = \frac{1}{5}$, so the sum of the series is $\dfrac{\frac{1}{5}}{1 - \frac{1}{5}} = \frac{1}{5} \cdot \frac{5}{4} = \frac{1}{4}$.

12.4 Taylor Series

One of the main ideas in this section is that a Taylor series is a function. Be sure to read the text on pages 735 and 738.

You need to know the general formula for a Taylor series at $x = 0$ and the specific formulas for e^x and $1/(1-x)$. Check with your instructor to see if you also need to memorize the Taylor series at $x = 0$ for $\cos x$ and $\sin x$. Study Examples 3, 5, and 6 carefully to get ideas for the solutions of Exercises 5-27.

1. We have

$$f(x) = \frac{1}{2x+3} = (2x+3)^{-1}, \qquad\qquad f(0) = \frac{1}{3},$$

$$f'(x) = -2(2x+3)^{-2}, \qquad\qquad f'(0) = \frac{-2}{9} = -\frac{2}{9},$$

$$f''(x) = 8(2x+3)^{-3}, \qquad\qquad f''(0) = \frac{8}{27},$$

$$f^{(3)}(x) = -48(2x+3)^{-4}, \qquad\qquad f^{(3)}(0) = \frac{-48}{81} = -\frac{16}{27},$$

$$f^{(4)}(x) = 384(2x+3)^{-5}, \qquad\qquad f^{(4)}(0) = \frac{384}{243} = \frac{128}{81}.$$

Therefore,

$$f(x) = \frac{1}{3} + \frac{-\frac{2}{9}}{1!}x + \frac{\frac{8}{27}}{2!}x^2 + \frac{-\frac{16}{27}}{3!}x^3 + \frac{\frac{128}{81}}{4!}x^4 + \cdots$$

$$= \frac{1}{3} - \frac{2}{9}x + \frac{4}{27}x^2 - \frac{8}{81}x^3 + \frac{16}{243}x^4 + \cdots$$

$$= \frac{1}{3} - \frac{2}{9}x + \frac{2^2}{3^3}x^2 - \frac{2^3}{3^4}x^3 + \frac{2^4}{3^5}x^4 + \cdots.$$

7. In the Taylor series at $x = 0$ for $\frac{1}{1-x}$, we replace x by $-x^2$ to obtain

$$\frac{1}{1-(-x^2)} = \frac{1}{1+x^2} = 1 - x^2 + x^4 - x^6 + \cdots.$$

13. In the Taylor series at $x = 0$ for e^x, we replace x by $-x$ to obtain

$$e^{-x} = 1 + (-x) + \frac{1}{2!}(-x)^2 + \frac{1}{3!}(-x)^3 + \frac{1}{4!}(-x)^4 - \cdots.$$

Hence

$$1 - e^{-x} = 1 - (1 - x + \frac{1}{2!}x^2 - \frac{1}{3!}x^3 + \frac{1}{4!}x^4 - \cdots)$$

$$= x - \frac{1}{2!}x^2 + \frac{1}{3!}x^3 - \frac{1}{4!}x^4 + \cdots.$$

19. In the Taylor series at $x = 0$ for $\cos x$, we replace x by $3x$ to obtain

$$\cos 3x = 1 - \frac{9}{2!}x^2 + \frac{81}{4!}x^4 - \frac{729}{6!}x^6 + \cdots.$$

We now differentiate both sides and divide by -3 to obtain

$$-3 \sin 3x = -\frac{18}{2!}x + \frac{4(81)}{4!}x^3 - \frac{6(729)}{6!}x^5 + \cdots.$$

$$\sin 3x = \frac{6}{2!}x - \frac{4(27)}{4!}x^3 + \frac{6(243)}{6!}x^5 - \cdots$$

$$= 3x - \frac{27}{3!}x^3 + \frac{243}{5!}x^5 - \cdots$$

$$= 3x - \frac{3^3}{3!}x^3 + \frac{3^5}{5!}x^5 - \cdots.$$

25. In the Taylor series expansion at $x = 0$ for $\frac{1}{\sqrt{1+x}}$, we replace x by $-x$ to obtain

$$\frac{1}{\sqrt{1-x}} = 1 + \frac{1}{2}x + \frac{1 \cdot 3}{2 \cdot 4}x^2 + \frac{1 \cdot 3 \cdot 5}{2 \cdot 4 \cdot 6}x^3.$$

31. If you are reading this before you work the exercise, stop and study Practice Problem 4. Then try the exercise again before reading on with the solution. We observe that the term in the series containing x^5 is $\dfrac{f^{(5)}(0)}{5!}x^5 = \dfrac{2}{5}x^5$. So $\dfrac{f^{(5)}(0)}{5!} = \dfrac{2}{5}$, and

$$f^{(5)}(0) = (5!)\frac{2}{5} = (4!)2 = 24 \cdot 2 = 48.$$

37. In the Taylor expansion at $x = 0$ for $\dfrac{1}{1-x}$, we replace x by $-x^3$ to obtain

$$\frac{1}{1+x^3} = 1 - x^3 + x^6 - x^9 + \ldots.$$

Integrating both sides, we have

$$\int \frac{1}{1+x^3}\,dx = \int (1 - x^3 + x^6 - x^9 + \ldots)\,dx$$

$$= [x - \frac{1}{4}x^4 + \frac{1}{7}x^7 - \frac{1}{10}x^{10} + \ldots] + C.$$

43. Since $e^x = 1 + x + \frac{1}{2}x^2 + \frac{1}{6}x^3 + \ldots$, for all x, we have $e^x > \frac{1}{6}x^3$ when $x > 0$. The idea now is to find a function which approaches 0 as $x \to \infty$, with the property that for every value of x greater than 0, this function is greater than the function $x^2 e^{-x}$. From the above inequality we have

$$\frac{1}{e^x} < \frac{6}{x^3}$$

and so

$$x^2 e^{-x} < x^2 \cdot \frac{6}{x^3} = \frac{6}{x}.$$

As $x \to \infty$, $\frac{6}{x}$ approaches 0. Since $\frac{6}{x} > x^2 e^{-x}$ for all $x > 0$, we conclude that $x^2 e^{-x}$ must approach 0 as $x \to \infty$.

12.5 Infinite Series and Probability Theory

This last section of the text combines the probability theory from Chapter 11 with the ideas in Section 12.3 (the geometric series) and Section 12.4 (the Taylor series at $x = 0$ for e^x).

1. If a coin is tossed 4 times, there are 16 possible outcomes:

    > heads heads heads heads;
    > heads heads heads tails;
    > heads heads tails heads; etc.

    This can be seen by observing that there are 2 possible outcomes on the first toss, 2 on the second, 2 on the third, and 2 on the four. Hence the total number of possible outcomes is $2 \cdot 2 \cdot 2 \cdot 2 = 2^4 = 16$. There is, however, only one way to toss 3 consecutive heads and then a tail. Thus the probability of this occurrence is 1/16.

7. The probability p_0 corresponds to a blue cab arriving first. Since there are twice as many red cabs as there are blue, one out of every three cabs is blue, so $p_0 = \frac{1}{3}$. To calculate p_1, we consider the possibility of one red cab going by before a blue appears. The probability that the red taxi passes first is $\frac{2}{3}$. The probability that a blue taxi passes is $\frac{1}{3}$. Hence the probability that first a red and then a blue taxi passes is $\frac{2}{3} \cdot \frac{1}{3} = \frac{2}{9} = p_1$. In a similar manner we deduce that $p_2 = \frac{2}{3} \cdot \frac{2}{3} \cdot \frac{1}{3} = \frac{4}{27}$. We conclude that $p_n = (\frac{2}{3})^n (\frac{1}{3}) = (\frac{1}{3})(\frac{2}{3})^n$.

13. The probability that a given page has at most four errors is $p_0 + p_1 + p_2 + p_3 + p_4$. We have

    $$p_0 = e^{-\lambda} = e^{-5} \approx .00674,$$

    $$p_1 = \lambda e^{-\lambda} = 5e^{-5} \approx .03370,$$

    $$p_2 = \frac{\lambda^2}{2} e^{-\lambda} = \frac{25}{2} e^{-5} \approx .08425,$$

    $$p_3 = \frac{\lambda^3}{6} e^{-\lambda} = \frac{125}{6} e^{-5} \approx .14042,$$

    $$p_4 = \frac{\lambda^4}{24} e^{-\lambda} = \frac{625}{24} e^{-5} \approx .17552.$$

Thus $$p_0 + p_1 + p_2 + p_3 + p_4 \approx .44063.$$

19. The probability that the number of claims is either one, two, or three is given by $p_1 + p_2 + p_3$. We have

$$p_1 = \lambda e^{-\lambda} = 10e^{-10} \approx .00045,$$

$$p_2 = \frac{\lambda^2}{2}e^{-\lambda} = 50e^{-10} \approx .00227,$$

$$p_3 = \frac{\lambda^3}{6}e^{-\lambda} = \frac{1000}{6}e^{-10} \approx .00757.$$

Thus $$p_1 + p_2 + p_3 \approx .01029.$$

25. The probability that x is an odd integer is given by
$$p_1 + p_3 + p_5 + p_7 + \cdots$$

$$= \lambda e^{-\lambda} + \frac{\lambda^3}{3!}e^{-\lambda} + \frac{\lambda^5}{5!}e^{-\lambda} + \frac{\lambda^7}{7!}e^{-\lambda} + \cdots$$

$$= e^{-\lambda}(\lambda + \frac{\lambda^3}{3!} + \frac{\lambda^5}{5!} + \frac{\lambda^7}{7!} + \cdots).$$

From Exercise 24 of Section 12.4, we find that

$$\sinh x = x + \frac{1}{3!}x^3 + \frac{1}{5!}x^5 + \frac{1}{7!}x^7 + \cdots.$$

Hence

$$\sinh \lambda = \lambda + \frac{\lambda^3}{3!} + \frac{\lambda^5}{5!} + \frac{\lambda^7}{7!} + \cdots.$$

This shows that

$$p_1 + p_3 + p_5 + p_7 + \cdots = e^{-\lambda}\sinh \lambda.$$

Chapter 12: Supplementary Exercises

1. $$f(x) = x(x+1)^{3/2}, \qquad\qquad f(0) = 0,$$

$$f'(x) = \frac{3}{2}x(x+1)^{1/2} + (x+1)^{3/2}, \qquad f'(0) = 1,$$

$$f''(x) = \frac{3}{4}x(x+1)^{-1/2} + \frac{3}{2}(x+1)^{1/2} + \frac{3}{2}(x+1)^{1/2}$$

$$= \frac{3}{4}x(x+1)^{-1/2} + 3(x+1)^{1/2}, \qquad f''(0) = 3.$$

Hence

$$p_2(x) = 0 + \frac{1}{1!}x + \frac{3}{2!}x^2$$
$$= x + \frac{3}{2}x^2.$$

7. $f(t) = -\ln(\cos 2t),$ $\qquad\qquad\qquad f(0) = -\ln(1) = 0,$

$f'(t) = \dfrac{2 \sin 2t}{\cos 2t},$ $\qquad\qquad\qquad f'(0) = 0,$

$f''(t) = \dfrac{\cos 2t(4 \cos 2t)+2 \sin 2t(2 \sin 2t)}{\cos^2 2t}$

$\qquad = \dfrac{4 \cos^2 2t + 4 \sin^2 2t}{\cos^2 2t}$

$\qquad = 4 + \dfrac{4 \sin^2 2t}{\cos^2 2t},$ $\qquad\qquad f''(0) = 4.$

Hence $p_2(t) = 0 + \frac{0}{1!}t + \frac{4}{2!}t^2 = 2t^2.$

The area under the graph of $y = f(t)$ is approximately equal to the area under the graph of $y = p_2(t)$. Thus

$$[\text{Area}] \approx \int_0^{1/2} 2t^2\, dt = \left.\frac{2}{3}t^3\right|_0^{1/2} = \frac{2}{3}\cdot\frac{1}{8} = \frac{1}{12}.$$

13. The series is a geometric series with $a = 1$ and $r = -\frac{3}{4}$. Since $\left|-\frac{3}{4}\right| < 1$, the series is convergent and converges to

$$\frac{1}{1+\frac{3}{4}} = \frac{4}{7}.$$

19. We observe that

$$e^x = 1 + \frac{1}{1!}x + \frac{1}{2!}x^2 + \frac{1}{3!}x^3 + \frac{1}{4!}x^4 + \cdots$$

Therefore

$$1 + 2 + \frac{2^2}{2!} + \frac{2^3}{3!} + \frac{2^4}{4!} + \cdots$$

$$= 1 + \frac{1}{1!}(2) + \frac{1}{2!}(2)^2 + \frac{1}{3!}(2)^3 + \frac{1}{4!}(2)^4$$

$$= e^2.$$

25. $\dfrac{1}{(1 - 3x)^2} = \dfrac{1}{3}\dfrac{d}{dx}\left[\dfrac{1}{1 - 3x}\right] = \dfrac{1}{3}\dfrac{d}{dx}\left[1 + 3x + 3^2x^2 + 3^3x^3 + \cdots\right]$

$$= 1 + 6x + 7x^2 + 4\cdot 3^3 x^3 + \cdots$$

31. (a) The fifth Taylor polynomial consists of all terms in the sixth Taylor polynomial with degree ≤ 5. Therefore, $p_5(x) = x^2$.

(b) Since the coefficient of x^3 in $p_6(x)$ is zero, $\dfrac{f^{(3)}(0)}{3!} = 0$. Therefore, $f^{(3)}(0) = 0$.

(c) $\displaystyle\int_0^1 \sin x^2\, dx \approx \int_0^1 (x^2 - \tfrac{1}{6}x^6)\, dx = \left. \dfrac{x^3}{3} - \dfrac{1}{42}x^7 \right|_0^1 = \dfrac{1}{3} - \dfrac{1}{42}$

$$\approx .3095 \quad \text{(exact value: .3103)}$$

37. $\displaystyle\sum_{k=1}^{\infty} 10{,}000\, e^{-.08k} = \sum_{k=1}^{\infty} 10{,}000(e^{-.08})^k = \dfrac{10{,}000(e^{-.08})}{1 - e^{-.08}}$

$$\approx \$120{,}066.70$$

43. $p_4 = \dfrac{4^4}{4!}e^{-4} = \dfrac{256}{24}e^{-4} \approx .19537$